THE

Electric Automobile

ITS CONSTRUCTION, CARE AND OPERATION

BY

C. E. WOODS, E.E., M.E.

PREFACE.

The many inquiries, both personal and written, made of the writer as to the operation, use and possibilities of the electric automobile have led him to write the following book on the subject; and as these inquiries have come very largely from people who are neither engineers nor mechanics, but are interested in the purchase and use of automobiles, the book has been written in a language as void as possible of technical nomenclature, although for lack of words to express some things, he has been obliged to use phrases commonly employed by technical and engineering fraternities. But the book is not meant in any sense as a technical or engineering work.

The effort has not been to produce a scientific treatise on the subject, nor even an elaborate composition, but to set forth in plain and unembellished English the facts and conditions surrounding the question in hand so that those interested in the subject might learn the elementary conditions and thereby purchase an electric automobile with some discrimination, and give it intelligent care and skillful operation. The writer trusts that the work will be received by the public in the spirit in which it has been written.

C. E. W.

The Electric Automobile

CHAPTER I

GENERAL CONDITIONS SURROUNDING THE INTRODUCTION AND USE OF AUTOMOBILES

The basic value of any proposition is its commercial aspect, and naturally the first query of inventor, capitalist and layman is: "What conditions exist that will make a market for automobiles or create a desire in the public mind for their use?" Some will say progression, the spirit of which surrounds us everywhere; others say, expediency and the desire for saving minutes and even seconds; others, again, their convenience and readiness for instant use; all of which are true but do not in a broad sense answer the question, but create another as to what has made all these things desirable on the part of the public as things necessary to its comfort and welfare.

The answer to this broader inquiry is that other means of transportation than by animal power, as by bicycle, trolley and train, have paved the way for the horseless vehicle; and so much is this so that, without reference to its novelty, the demand is now seriously of a business nature in both the commercial and domestic world, and especially as a means of conveyance for the public at large, by cab, cabriolet and coupé.

At one time, the horse furnished the power used in all the manufactures of the world, and no greater honor could have been awarded him than to adopt his muscular efforts as the standard of measurement when the steam engine was invented. It certainly was with no disparagement of the horse that the steam engine was everywhere preferred for furnishing power for factory use. The reasons for this are too numerous to give, and if enough could not be found, the horse himself would cheerfully furnish any number that might be lacking. The horse has done his best for man. He has tugged and pulled everything from canal boats to street cars until, from time to time, relief by some mechanical means has eased his work; and no matter where the displacement has occurred, it has been at once evident beyond the chance of argument that such mechanical device, no matter how crude or imperfect it may have been, rendered a far superior service and at an immense saving in cost. What board of directors of a street railway company would to-day recommend building a street railway system to be equipped with horse power? They would simply be considered unsafe men to guide the affairs of any company.

When we review all that has been done by mechanical devices toward the displacement of animal power, it is very hard to refrain from drawing a conclusion that the horse must go; that is, speaking in the broad sense of the word. Mechanically propelled vehicles for all purposes are here. This applies to our railroads, our street cars, our steamships and every means of conveyance for which a vehicle is needed, as well as to the automobile, which is the last in the list to be grasped by inventive genius as a means for mechanical conveyance on our streets and boulevards.

"Last in the list" is wrong. It is more like a case in which first is last and last is first. Few observers seem to know that the magnificent schemes of transportation embodied in our present railroad systems had their origin in the horseless or motor carriage, but such is the fact nevertheless. An illustration of this is given in Plate II, which shows the beginning of mechanical transportation —the combination of the Locomotive and Coach in its first conception and a type of vehicle made in 1833 known as the Squire Steam Carriage, a four-wheeled vehicle which carried its passengers in front and its propelling power behind. The separation between the motive power and the passenger vehicle has been preserved down to the present day in railroads.

Another very interesting motor vehicle, illustrated in Plate I, was described in the Mechanic's Magazine, of January 1, 1834. This vehicle was designed and built by a Dr. Church in England, and was also propelled by steam. It was intended to carry forty to fifty passengers.

It is stated beyond all question that to a Frenchman by the name of Nicholas Joseph Cugnot belongs the honor of constructing the first mechanically propelled vehicle ever used. It was propelled by steam and was built and run by him between the years 1763 and 1769, so that we are safe in saying that the advent of the horseless vehicle was at that period. These vehicles were first intended to operate in roadways, but the crude and cumbersome condition of their mechanical appliances, the heavy losses by friction in their working parts, and their inefficiency as

4

a whole compelled inventors to seek some means of increasing the working capacity.

The first discovery was that a good roadbed was necessary; and also, as the appearance of the vehicles was a great disturbance in the streets to both residents and horses, an isolated road was necessary for them to operate upon. To improve these conditions so as to lend successful operation to the early inventions, roads were built, at first with tracks made of timber and called rails with the word "road" affixed, a name that has held from that day to this. By this, inventive genius grasped the horseless carriage or motor vehicle as one of its very first propositions in mechanical transportation, and has only taken it up again in its original form when the necessary refinement of mechanical conceptions and the construction of roads and streets have made it practicable to do so.

All this is not said with any feeling of delight in helping to displace the horse—this most useful animal, who has served us so long and faithfully—but rather in deference to the eternal economy of things in this progressive age.

Time has now proven that vehicles can be propelled mechanically at much less cost, with greater rapidity and less noise and clatter than by animal power, and the great question which has been worrying taxpayers for centuries—that of street paving and street cleaning—is settled at once, absolutely and satisfactorily, for all time to come, and for all people concerned, when the new condition of affairs shall have become general enough to be understood.

A horse must be waited on; he needs constant and special attention; his endurance is extremely limited; after a comparatively short service he must be stabled, groomed, fed, and curried, and a man must be employed to do this. New paving blocks must be purchased and put down in the street where his iron shoes have worn or displaced the others, and another man must be employed to do this. A man must be employed to remove his excrements from the streets. He is subject to colds and almost as many diseases as man himself is, and at best his life is very short; and again, when the immediate necessity of his service is suspended, there is no suspension of the expense attached to keeping him. This always goes on at an undiminished rate until he dies.

On the other hand, the electric vehicle starts out and is capable of covering three times the distance in the same length of time. It is fed (i. e., charged) at a fraction of the expense and with comparatively no attention. There is practically no depreciation of the pavement and roadways by contact with its rubber tires. Well paved streets should last for an indefinite period of time with its exclusive use of them. There is no dirt and refuse to be cleaned from the street, and no dust is thrown into the rider's eyes by its operation. No expense is entailed, except the cost of investment, when not in service. Noiseless and odorless in its operation, it has at the same time the same flexibility of power application as is exercised by the horse, if not a greater one.

After all, it is but the old story of mechanics displacing muscle, with the exception that in this case the displacement will be on a much larger scale than anything of that nature which has gone before.

6

Whatever the horse can do can be done better, more quickly and cheaply by the automobile, whether it be on hurried errands of business by cab or coupé, hauling of merchandise over the streets, or pleasure driving along smooth boulevards.

Such, in general, are the prior and present conditions of things that have gradually led up to the introduction of the automobile to the general public and its acceptance thereof. And the preference for electric vehicles, especially those of a public nature, will be exactly in the same ratio as the preference for trolley cars over cars drawn by horses, for there is the same relative condition existing between the two regarding expediency, and in addition there is a far greater superiority in refinement and splendor attached to the electric vehicle. While some of its first usages may be simply from curiosity, this will quickly lead into custom.

The next question of interest and vital importance is, to know by what means the electric vehicle can be made to work more economically and more advantageously than the horse-drawn vehicle, and two or three very terse and simple comparisons will be drawn to illustrate this point.

Taking, first, the ordinary cabs, as drawn by horses, and based on the average rates established by all liveries of fifty cents for the first mile and twenty-five cents for each additional mile, the horse vehicle will cover about five miles in one hour and is good for about four hours' work per day, making a possible twenty miles and six dollars revenue as an average. The same cab, to make further earnings or mileage,

7

must have another horse attached to it, twenty miles per day being all that one horse can possibly do. Short trips in the centers of large cities will sometimes run the earnings up to seven or eight dollars per day.

Now let us see what the electric cab can do. An average speed of nine miles per hour would give it thirty-six miles in four hours. It would earn in that time, on a mileage basis of fifty cents for the first mile and twenty-five cents for each additional mile, two dollars and fifty cents per hour, or ten dollars for four hours, at a cost not to exceed two cents per mile for electricity, or seventy-two cents, and twenty-five cents an hour for the driver, or one dollar, making a total cost of one dollar and seventy-two cents for the operating expense of running only.

The horse cab will have earned in the same time, on twenty miles, six dollars, at a cost of eighty cents for the horse and one dollar for the driver, figuring the latter on the same basis of twenty-five cents per hour. Thus the electric cab has, in the same length of time, sixty-six and two-thirds per cent greater earning power, and at about three and one-half per cent less cost.

Now let us look at other conditions. The first cost of an electric cab is $3,000, as against $1,200 for a good cab, horse and harness, or almost three times as much. The depreciation of the latter is about thirty-three and one-third per cent, while the depreciation of the former is not to exceed twenty per cent. But we have not as yet by any means reached the earning capacity of an electric cab per day. After using it, say, four hours in the morning, we can charge, say, from 12 noon until 1:30 in the afternoon, and then have until 5:30 to make another thirty-six miles.

Then, by charging from 5:30 until 7, the cab has thirty-six miles more for its night's work. Further comparison here would be superfluous so far as cabs go.

In addition to this, the writer gives some figures as pertaining to the cost of operation of a gentleman's private carriage, like a Stanhope, as compared with the horse-drawn vehicle of the same character:

COST OF HORSE VEHICLE

Cost of Stanhope. . . . .$500.00 .
Cost of horse. . . . . 250.00 .
Cost of harness. 100.00
Total . . . .$850.00 .

COST OF ELECTRIC VEHICLE

One Stanhope. . . . . $2000.00.

Cost of electric vehicle to run twenty miles a day, twenty cents. Cost of stabling, eight dollars per month. Cost of current per month, six dollars. Total cost to operate, fourteen dollars per month.

Cost of stabling and feeding a horse, twenty-five dollars per month.

Daily mileage obtainable with one electric vehicle, eighty miles.

Daily mileage obtainable with one horse vehicle, twenty miles.

Same mileage with horse-drawn vehicle would require four horses at $250 each, or $1,000, and $600.00 for carriage and harness would make a total cost of $1,600.

Figured for a year, this would be as follows, taking sixty miles a day as a possible average family riding: II

ELECTRIC VEHICLE

	per annum
Current, $18.00 per month.	$216.00
Rent, 8.00 " ".	.96.00
Total for maximum work.	$312.00
20 % depreciation.	.400.00
Making a total cost of.	.$712.00 .

SAME MILEAGE BY HORSE VEHICLE

	per annum
3 Horses per month, $75 each.	$.900.00
Depreciation, 20%.	.170.00
Or a total of.	$1070.00.

And two men would be required to take care of the horses, where only one, and that a general house servant, would be required to take care of the electric carriage—a further saving of at least $600 per annum. Some men say that they can board a horse for less than twenty-five dollars per month. True! But after shoeing and veterinary surgeon bills are paid, the average cost of keeping a horse and keeping him properly is never less than twenty-five dollars per month, notwithstanding whatever may be said to the contrary.

In both of these sets of data, no reference is made to what is known as a battery exchange; that is, taking a discharged set of batteries out of a vehicle and substituting for them a duplicate set that is charged.

While this method of procedure is necessary, perhaps, with some forms of battery, it gives the writer pleasure to say that he is at the present time working with a battery that, for all practical purposes, can be charged in one hour's time, or at the outside, one hour and fifteen minutes. This eliminates all the difficulties that have beset the use of an electric vehicle by its limitation on mileage capacity and the length of time which heretofore has been required in the charging of batteries.

To illustrate this more forcibly: The writer has taken a small road buggy and ridden twenty-five miles from 9 o'clock until 12 in the morning; charged it until half-past one; ridden until half -past four another twenty-five miles; charged it until 6:30 and ridden another twenty-five miles before 10 o'clock in the evening, which makes a total day's work with one of these little buggies of seventy-five miles. Surely such conditions as these make the use of an electric vehicle not only practicable in every sense of the word, but applicable to any and all classes of street transportation.

The cost of running this buggy, based on prices for current as given by electric-light companies, is about sixty cents for seventy-five miles. The mileage capacity of one of the buggies referred to is nominally twenty-five miles, although they have run twenty-eight and thirty miles on frequent tests that have been made. Larger vehicles run thirty-five to forty miles on one charge of the batteries. Tests recently made on a Stanhope gave a mileage, when every condition was most favorable in road and weather, of fifty-seven miles on one charge of the batteries.

Manufacturers of electric automobiles are often asked the question: "Will not a battery sometime be made to run a vehicle one hundred or more miles on one charge?" While this may be possible, the writer hardly thinks it will ever be probable, and his experience with the public is that it is wholly unnecessary, speaking generally and from a practical point of view. Improvements now being made in batteries, in the attempt to make them stand more rapid charges and discharges are much more important than one which will give them a larger mileage capacity. It is also necessary to supply conveniences for making these charges by the public lighting companies in every city. This is said because from a pleasure rider's point of view, after a man has been riding on the streets and boulevards for three or four consecutive hours, he is tired out and the pleasure of further riding is completely dissipated. For commercial purposes, fifty to sixty miles per day with a delivery wagon is all that two men can possibly cover and make their stops and deliver their packages, as something like fifty per cent of the time between leaving the shipping department and returning to it is consumed in making stops and deliveries.

The question is also often asked: "Will not the weight be greatly reduced in batteries?" This is hardly desirable beyond a certain point. Automobiles are a mechanical proposition of a self-driving nature, and for the amount of power applied to the driving wheels a certain weight is absolutely necessary to give a certain traction. Then, while a vehicle may require but two horsepower to drive it at a given speed on a level roadway, deep mud, heavy grades and bad places to start will often call for five or six horse-power to drive the same vehicle, and weight enough must be had to give traction for that power. The present storage battery combined with the vehicle complete weighs only about twenty-five per cent in excess of the required weight,—the battery itself being from forty to fifty per cent of the weight of any well-constructed and designed electric vehicle. It must be remembered that a reduction of twenty-five per cent in weight would not make a corresponding reduction in power required—all of which is merely a mathematical proposition.

The simplicity attached to the operation of an electric vehicle by any person of ordinary intelligence is too well known to need comment at this point; but it is found from experience that there is the same difference in the care taken of an electric vehicle that there is among men who attend dynamos and steam engines, or drive horses, with a corresponding difference in troubles and aggravations.

The writer knows of vehicles that have never given any trouble after six and eight months of hard use, and of other vehicles that the owner is bothered with from sheer carelessness and negligence every time he goes out. This rule holds good throughout all mechanical, electrical and other propositions of that nature. In nine cases out of ten where trouble has been reported, it has been found to be the fault of the owner or driver. First, he will not keep the solution in the batteries up to the proper height—a thing that should be attended to at least every two weeks. Again, he will start out many times with the batteries far from fully charged.

It is to be regretted that no one as yet has invented an instrument which is in any sense of the word accurate for measuring the charged or discharged condition of the batteries while in use on a vehicle. The drop of potential method has been employed almost exclusively and is accurate in itself. The trouble has been to get an instrument made so that it is sensitive enough to be accurate and at the same time durable enough to keep its accuracy when bumped around on the streets.

Other owners of automobiles will never stop to put a little oil on their bearings and seem to think because they are ball bearings and it is an automobile, it can run forever without being looked after. There are those, also, who show no consideration for a vehicle in driving it over rough streets, street crossings, and car tracks. If a purchaser of an automobile would give it the same careful consideration and treatment in its usage that he would give a fine horse and carriage, which costs him something like an equal amount, his troubles would disappear like snow before a summer sun.

The question of the durability of batteries has been one of vital importance and the experience of a great many manufacturers has been a most bitter one in this respect. The excessive charge and discharge demanded in this work, as compared with ordinary battery work, the rough usage received on the streets and the negligence and carelessness of both drivers and owners have been the problems that stared the automobile manufacturers in the face.

In this connection, the writer can only cite his own experiences, as other manufacturers are very careful in their statements about what their batteries will actually do. The batteries used by the writer at the present time have a life, with good care and intelligent looking after, of from twelve thousand to fifteen thousand miles before it becomes necessary to renew the positive plates, which are about one-third of the original cost of the batteries. He has tried some makes of batteries in which everything desirable was found as regards mileage, etc., but upon test they would not run more than a thousand or fifteen hundred miles before the positive plates would be almost entirely disintegrated. Nine thousand to ten thousand miles is an average year's work for a battery except when it is used in a public conveyance; and so far satisfied is the writer now with the battery proposition that he will insure the maintenance of batteries for so much per annum for a period of years, which settles once for all any doubts that may be in the purchaser's mind.

Another question which is asked nearly every day is: "Will not the cost of electric automobiles be greatly reduced in a few years, in the same degree that bicycles, sewing machines, and other mechanical devices in that line have been subject to?" The reply is invariably: "No!" While the cost may be reduced a slight percentage from what it is now, it should be borne in mind that the art of carriage building is an old art.

This also applies to wheels, ball bearings and axles, and so on for the various component parts that comprise an electric vehicle. The art of building electric motors and other accessories of that nature is already an old art. No labor-saving machinery, to any extent, can be devised in addition to that already in existence for reducing the cost of manufacture of the parts enumerated. The cost of the storage battery may be reduced somewhat, but the percentage in that alone would not exceed twenty per cent, and the battery is about twenty per cent of the cost of the complete vehicle. The only possible way that the cost of production of these vehicles can be reduced is by a well-disciplined and systematized method of manufacturing them, and a volume of business large enough to establish the necessary system and discipline.

Again, the automobile is not a single item produced like the bicycle and sewing-machine. It has such a wide range of application in its utility and design, such a wide range of prices in its cost of production and marketing, such a multitude of different elements to cater to by the various demands for its usage, that the question becomes entirely foreign to any manufacturing proposition as compared with a bicycle or sewing-machine.

A first-class Brougham or Victoria, as made by the best carriage makers in the country, can be purchased no cheaper to-day than it could ten or fifteen years ago. There are some lines of light buggies and delivery wagons which can be bought cheaper now than fifteen years ago, but automobiles are subject to the same process of manufacture and by the same improved machinery that these vehicles are.

Again, it should be remembered that the carriage part of the vehicle, that is, body, painting and trimming, is but a small part of its total cost. It is necessary to use ball-bearing axles, and they are very expensive to construct. This is required because every friction-reducing device possible must be used. It is necessary to use rubber tires in large sizes, first, to preserve the motors and batteries; and, second, to give the necessary traction. Steel tires can never be made to operate successfully on motor vehicles. Then come the electric motors, storage batteries, controllers and other devices, all of which are, in themselves, expensive as compared with the ordinary vehicle, and have to be treated as such in the selling price. While the first cost of electric automobiles is in many cases in excess of the horse-drawn vehicle, two, three and four times over, its utility is also in the same ratio.

In the writer's work with motor vehicles, he has entirely abandoned the use of wire wheels, pneumatic tires and tubular construction of all kinds. The pneumatic tire was designed for use on a bicycle in order to give some spring to a vehicle which otherwise had none and make it easier to operate. But this admirable design contained in its prime principle no relation to motor vehicles, which, when properly suspended on adequate and well-adjusted springs, cannot be improved in the least by the use of pneumatic tires, especially when the tire has to be pumped up to the extent of a hundred and twenty-five pounds to the square inch to keep it from deflating too much and thereby consuming too great an amount of energy. Wire wheels on bicycles have proven satisfactory in their construction and operation because there the load is hung centrally over the wheels and equally balanced on either side. In the motor vehicle the load is hung between two wheels, and the strength and durability of the wire wheels are taxed in the wrong direction. In very light vehicles they could be made strong enough, but their appearance as a carriage production, from a carriage user's point of view, is prohibitive. The writer has abandoned tubular construction because a carriage should be made in all of its essential parts so that it can be repaired by any carriage manufacturer in any city; and the bolts, nuts and carriage iron work on the vehicle must be of this nature and made to this end. The vehicle is too expensive in transportation to have to be sent back to the factory whenever repairs are necessary. Again, the owner of an automobile equipped with pneumatic tires would present a rather strange appearance carrying his automobile on his shoulder to some automobile hospital for the purpose of getting a puncture repaired, as can be done with a bicycle, because he would be performing an impossibility. The facts are, however, more tires blow up than are punctured, owing to the weight carried and the varying range of pressure in the tire as it gradually leaks out; which action, in time, disintegrates the rubber and shortens the life of the tire. Pumped to something like one hundred and twenty-five pounds to the square inch, it is as hard to ride on as a solid tire, and consumes about the same power to run the vehicle; but it requires more power as it gradually deflates.

In a book of this nature, the full scope of the automobile industry and its future possibilities cannot be described. One might go on and speculate concerning the future of the automobile to such an extent that he would be called visionary and still be strictly within the confines of the truth. Roads might be anticipated, built from one city to another, of macadam construction, on which nothing but rubber tired vehicles without horses would be permitted to operate. The expense of construction and maintenance of such a road would be so small that it might easily come within state and county administration. There are many other things that will well up in the mind of one who constantly studies possibilities of the automobile industry that would not be of interest here.

The passing of the age of the horse has not come, as many suppose, with the introduction of the automobile. It began with the first introduction of mechanics into commercial and industrial life. And the automobile is the fourth or fifth step in this ladder of progression. The change will not be instantaneous, but as gradual as all other great changes have been. Step by step, in an ever-increasing advance which multiplies itself as it goes on in true differential order. The law of nature is from homogeneous to heterogeneous, and this law is as immutable in the conditions, affairs and usages of man as it is in things. When we review the mechanical progress of the last hundred, or even fifty, years, and compare the small development in mechanics that took place in all the centuries that passed before, we cease to wonder at what the next century will bring forth in the concentration of our inquiry as to the morrow.

CHAPTER II

CARRIAGE CONSTRUCTION, DESIGN AND SPEED IN CONNECTION WITH AUTOMOBILES

The name "Automobile" is misleading from one point of view at least, i. e., as regards the carriage user's and carriage manufacturer's requirement for different names for different styles of carriages, as a single name applied to so broad a proposition is an error inasmuch as it savors of a name applied to some single-item product, as like unto a bicycle, which primary name is never used in connection with a tandem, quadrant, sextant, and so on. One never thinks of saying "bicycle tandem," and it is somewhat of an imposition to expect people to say "Automobile Stanhope," " Automobile Cab," and so on, through all the names of different styles in carriage design. The name in itself is significant as distinguishing a power-driven vehicle from a horse-drawn vehicle, but it is distinguishing in this respect only and is never used in connection with a carriage when it is desired to name some specific style. Neither does the word "Electromobile," which has been suggested by some of the electrical fraternity, make it any better. Each different style or character of vehicle has had its own specific name for years, and when we go to market with the self-propelled vehicle, we invariably find people talking in the same phraseology that they have been accustomed to and making inquiry for Stanhopes, Runabouts, Broughams, and so on through the whole category of names, usually prefixing the word "Electric." The writer's personal experience with the public is that, from sheer force of custom, they have gone into citing vehicles propelled by electricity as electric Stanhopes, electric cabs, etc., without allowing the name "Automobile" to enter into the commercial side of the question, except as a general name for the whole product manufactured.

To show how this is, let one man say to another that he has an automobile and he expresses nothing to the listener as to the character of the vehicle he may have and is invariably asked what this is. He then answers that it is an electric Stanhope, buggy, Brougham, or whatever class of vehicle he may possess. If only one style of vehicle were manufactured, "automobile" could then be used in the same sense that "bicycle" has always been used, and distinguished in the same way in improvements by the time period of its model. Ultimately, there is hardly any doubt but that the electric vehicles will be called by the same names that have been given the same character of vehicles drawn by horses, with the word "electric" prefixed when necessary.

There is a mistaken idea with many people, who have not given the subject any thought, that automobiles are sold for their novelty and because they go without a horse. But this is wrong. The purchasing public, which uses automobiles, buys them primarily for the same purpose for which it has always purchased any class of vehicle, namely, because a carriage or a vehicle is needed for personal transportation, convenience and comfort; and as it is among the better class of carriage users that automobiles are generally sold, they demand the same diversity of design, the same elegance in finish, the same magnificence in appointment, and the same easy riding qualities that they have always been accustomed to when drawn by horses. Consequently, the prime object of their purchase being to secure a fine carriage, the appointments of which they are thoroughly versed in, the automobile proposition is just as much of a carriage-manufacturing proposition as it is a mechanical or electrical one.

This means a great deal, for a successful system of propulsion must be designed that can accommodate itself, generally speaking, to all of the various styles of carriages with the use of which the public is familiar.

There is also a very general impression prevalent that an automobile to carry two persons ought to be sold at a price but little in excess of the price of a good horse and buggy. The great economy of its operation, the skilled workmanship and first-class materials that must go into its construction if it is to stand the various conditions of road service at speeds much in excess of the horse it is intended to displace, and the greater mileage it is capable of making in the same time, are lost sight of, or rather are never thought of at all; and it is well to say here that those who expect to get a motor vehicle of this character at the same price for which they can purchase a horse and buggy, and consider it from that point of view only, are doomed to disappointment. The effort to produce a motor vehicle of this kind by some manufacturers has resulted in disaster to the vehicle, to the purchaser, and to the manufacturer himself. They have contended that they ought to give the

22

public what it wants, and to economize have built a vehicle altogether too light for the service required of it, and by the elimination of economical features of operation, like ball bearings and such other refinements as are necessary to save power, have only produced a harshly-operating and unfinished-appearing vehicle.

The unsightly appearance of automobiles has been commented upon in this country a great deal. The trouble has usually been that engineers, electricians and mechanicians have been the original authors of the automobile, and their minds have been so concentrated upon the development and perfection of the mechanical and electrical parts that they have entirely ignored the artistic side of it. This was undoubtedly brought about by the indifference and skepticism, as well as opposition, offered the advancement of the motor vehicle from legitimate carriage manufacturers themselves, to whom such men refrained from going for advice. There is no question but that this problem belonged to the carriage manufacturers, and had they taken hold of it in time, they would have preserved to themselves an industry which they rightly had earned by prior experience and conceptions as carriage producers. But the period for that has passed, as also the period for crudeness in appearance has passed.

To-day carriages are built in practically as many varieties of style as is the ordinary horse-drawn vehicle, and have just as elegant an appearance, if not more so in many respects, as the automobile has brought with it some new creations in carriage design which were impossible with a horse in front of the vehicle.

The writer is often asked: "Is not something entirely new to be created that is quite different from a horse-drawn carriage in the same sense that a bicycle was— a new vehicle creation in itself?" This is not possible as a general proposition, and would only apply to some extreme novelty. We must have four wheels; we must have springs; we must have a seat to sit on and a body on which to build the seat; we must have a top of some kind for various vehicles; and to make a carriage that would run and eliminate a part or all of these features is simply an impossibility. Many attempts have been made, both here and abroad, to construct novelties of this kind, usually in a three-wheeled vehicle, the power being sometimes applied to the single wheel, as it saves the cost of a differential gear, cost of one wheel, makes the vehicle possibly lighter and at the same time lowers the price. But this design has invariably resulted in a sacrifice of stability, as a single driving wheel has not sufficient traction and has a tendency to be slow on greasy pavements and rough streets. In addition to this it makes an awkward-looking vehicle and is not susceptible of anything like general carriage design. First of all, we must have a fine carriage production. Next we must have a power application in the simplest way possible. And, third, we must have a power so universal in its supply and so simple in its manipulation that it can be put with impunity into the hands of any and all who wish to own an automobile; and to-day the electric vehicle has been shown to possess all of these qualifications.

Another great query is: "Will the electric vehicle go through snow, and mud, on country roads, etc.?" The broadest reply that we can make to this is that it will go anywhere that one could reasonably expect to drive a horse. An interesting article on this point appeared in an issue of the Horseless Age, of December, 1898. Electric station owners are now making a very important movement to popularize the electric vehicle by providing their plants with the apparatus necessary to charge a carriage, and this movement is doing more to invite public confidence in the use of the electric vehicle than any other single step that has ever been taken. Experiments are being made with an electric hydrant, as it is called, which will be situated at various streets and junctions where a proper meter system will be installed and where the owner of an electric automobile can stop, drop his ten cents into the slot and get his supply of current in a comparatively few moments' time. When such systems as this have been introduced in the cities, and extended out on to the country roads between one city and another, the electric vehicle may make any distance and travel on any journey; and this will come in time.

Perhaps no stronger argument was ever used against the general introduction of railroads throughout the country than that locomotives had such a limitation of fuel and water supply contained within themselves they could not go far, and that to transport fuel from one end of a State or country to another to supply the locomotives would be so expensive as to make them impractical.

How speedily and with what success this limitation was overcome it is needless to mention. We take a train in New York and ride through to San Francisco, and the question of being blocked between towns for lack of fuel or water is something that probably never enters a single passenger's head at the present day. And so, in its own sphere, will the day come when the electric vehicle will operate in the same manner.

In designing an automobile, the question confronts him who would manufacture it as to what form of wheels, what form of tire and what form of ball bearing, etc., it is most desirable and necessary to use. There has been a great deal of dispute as regards rubber tires for electric vehicles, for, in one form or another, rubber tires must be used. A notable example in this direction is presented by Mr. J. Herbert Condict in the Horseless Age of March, 1899, in which he says:

"The tire situation is at present the absorbing subject of our day thoughts and night dreams. Were the streets in this great metropolis paved in an up-to-date manner there would be comparatively little cause for anxiety on this score, but, with the antiquated and despicable cobble, the vase is very different. Over twenty distinct types of tires have been tried, or are to be tried in the near future, and others will probably follow. Solid, single and double tube, pneumatic, cushion, clincher, sectional, protected and unprotected have all had, or are to have, a trial. Their faults are many, their virtues few, and we are still on the search. We have heard of some eminent authorities in the motor-vehicle field who say that the solid is the only tire. They are rendering judgment without full knowledge of the facts, and from a purely local standpoint. They are evidently not acquainted with the streets of Manhattan. A very satisfactory arrangement for Broughams is that of rear pneumatics and front solids.

"The wheels also require most serious consideration. The severe strains unavoidable in crossing railroad tracks and other inequalities, and running against curbstones, speedily put out of service even the most substantial constructions. Everything, from the light and airy bicycle construction to the dishpan wheels now so familiar on our streets, has been tried, and still there are more to follow. I mention these few items simply to indicate some of the most important and particular directions in which the work is progressing."

The writer's own experience has been very different in its results from that indicated by Mr. Condict After the construction of a few vehicles, early in his development of them, on which he used pneumatic tires and went through the same experience indicated by Mr. Condict's article, he adopted the hard or solid rubber exclusively, and designed diameters of wheels, width of felloes, etc., to accommodate such sizes of tires as by experience proved best suited to the many and different styles of vehicles to be built, for he had discovered that the resiliency of pneumatic tires was entirely lost when the carriage was properly designed and the weights properly distributed on its points of support and the latter placed on properly designed springs. The easy-riding carriage for any purpose depends entirely upon its springs for this qualification, and there is no reason why the automobile, with its heavier weight, should be any exception to the general rule. If, however, carriage design embodies the placing of a set of batteries over one set of springs, making a very unequal distribution of the load—which in itself is always a faulty design—it cannot be expected to be easy, and a very large and not too much inflated pneumatic tire may help the difficulty a little; but even then when tires are inflated to the pressure necessary to give an economical power effort, there is scarcely any more resiliency left in them than that given by the hard rubber tire, and their unsightly and objectionable appearance as applied to a general carriage production, is too well known to need comment here. Wire wheels, from a carriage point of view, are objectionable, as was cited in our first chapter, and are still further objectionable when repairs are considered. If the carriage manufacturer has not tried to produce an automobile, it is fair to say he will make a very vigorous effort to get all the repair work he can and the general manufacturer of automobile carriages should have this in view as well as other things. The wood wheel has stood the test of years. Its construction is well known by some twenty thousand carriage makers in the United States; also its necessary repairs; there is no crystallizing of wire spokes from unequal strain or adjustment; there is no crushing of the rim from running the vehicle on a deflated tire that accompanies the use of wire wheels and pneumatic tires, which cannot be repaired by ordinary carriage makers. All these things are entirely eliminated when the wood wheel and hard rubber tire of proper construction and design are used. The wood wheel and hard rubber tire are strictly within the requirements of an automobile purchaser who is, and has been, a user of fine carriages. The writer knows of vehicles equipped with wood wheels and hard rubber tires that have made 15,000 to 18,000 miles without ever having been up for repairs, and certainly this is a practical proposition beyond all peradventure of experiment. From this it is easy to gather that when the carriage is properly designed and constructed, it makes little or no difference in its riding quality whether pneumatic or hard rubber tires are used, and as the latter are more reliable and less expensive, common sense at least should give them preference. These things are not theories, but are actual experiences duly recorded.

Another point of construction is bicycle tubing or tubing of that nature for frame work or running gears,—in other words, bicycle construction for supporting the carriage and its weight as compared with regular and well-known carriage methods of construction. Tubing can, without doubt, be made strong enough, but that is not the question altogether. We must have the entire carriage construction in such shape that it can be repaired by the same class of artisans, blacksmiths, etc., that is now employed by the carriage makers throughout the country. The objections to the use of tubular construction are its method of general construction, being brazed together at many points, and the fact that tubing once bent or damaged can never be made serviceable or satisfactory again. Frequently, therefore, new tubing must, perforce, be brazed and put into place, and, as it is a material not understood by carriage manufacturers, its use compels the purchaser to rely upon the factory for such repairs, however far the factory may be situated from him. A motor vehicle should be constructed in all of its iron work, its running gear and axles, the method of putting on its springs, etc., as nearly as possible after the methods now in existence in the carriage world, using, as far as practicable throughout the vehicle, standard carriage hardware. In this way the purchaser of an automobile has a resource at his own door for such repairs as he may need from year to year in addition to his regular painting, varnishing and trimming repairs.

An article on this point appeared in the Automobile Magazine of November, 1899, which was one of the most distorted statements the writer has come across in some time. No metallic work or construction by tubing of metal could be applied to carriage bodies in any general sense of the word as covering the many different styles demanded; and as the amount of tubing that could be successfully used in a carriage body for supporting the body or binding the running gear together would not lighten a cab as illustrated in the article to exceed two hundred pounds, the writer fails to see any logic in making the statement that by making the frame and body of tubing at least 1,200 pounds could be saved in weight. The figures given in the article, as of a cab weighing 4,900 pounds, are wholly erroneous, as the best makes of Hansom Cabs at the present day weigh as follows: Batteries, 1,100 pounds, with the rest of the vehicle weighing but 1,600 pounds, or a total of 2,700 pounds complete for the vehicle. If the entire running gear were made of tubing, not more than 200 pounds in weight could possibly be saved.

Ball bearings for axles and other moving parts of an automobile are almost as diversified in their construction and application as are its other important elements, but have proven successful in their every requirement, notwithstanding the heavy loads and high speed tasks which have been imposed upon them. One valuable feature in conjunction with the ball bearing axles is usually lost sight of and its whole value is cited on its easy -running or friction reducing qualities. This is that it needs so little attention in its demand for lubrication. Every two months is often enough to supply lubrication to a well-designed and well-protected ball bearing axle. There has been a great deal of controversy as to whether a three-point bearing or a two-point bearing is the most successful. This will depend largely on the number of rows of balls used, the treatment of the metal in hardening, and the general conditions created under which they must labor. The difference in amount of friction saved in either case is almost negligible, though it would probably be in favor of the three-point bearing, but its tendency to cone cutting is much greater on account of the more minute contact of the balls. The two-point bearing has been used by the writer exclusively in his work for the last eighteen months, and out of several hundred vehicles constructed during that time, the first set of cones remains yet to be sent in for repairs, which, in itself, tells the story better than a whole volume of theory.

Within the last few weeks some very remarkable tests have been shown on roller bearings, and some new designs recently created from two or three different sources give promise of great results, both in friction saving and in the trivial amount of care they require. It has been shown that friction is reduced one-half as compared with the ball bearing, and that the roller bearings in vehicles, after once lubricated, need no further attention for months. All this, however, remains yet to be proven by actual time tests.

The question of noise in a vehicle is something that has been reduced so that in a well-made vehicle it is scarcely perceptible even to the occupants of the vehicle. Were a motor vehicle to make as much noise as a horse's feet on the street, it would not be tolerated for a moment. One of the tritest remarks ever made on this subject is found among the editorials of the April, 1898, number of the Horseless Age, which says:

"The anti-noise society organized some time ago in New York ought to espouse the cause of the motor vehicle and good roads. Horses and the cobble-stone pavements they require for a foothold are two of the most prolific sources of noise in the metropolis."

We now come to the question of the rights and the speed of the automobile on the streets. The first point is the necessity of good judgment and moderation of speed. High speed is perfectly safe under certain conditions, and it is assuredly very exhilarating to ride in an automobile at a swift pace when the place and conditions of the street permit. But we must not forget that we still have the horse with his vehicle attached to him, and his defenders with us on the streets, and that, to a certain extent, we must conform to the general laws governing these vehicles. The automobile is a road vehicle, and where roads are good and few vehicles are encountered, high speeds are permissible; but where there are many vehicles to share the common rights of the road, majority always rules and the operator of an automobile must invariably conform to this rule, both for his own safety and for the safety of those who are not so well able to take care of themselves as he is.

The natural inquiry is: "What is the limitation of speed that can safely be used on city boulevards and streets?" As motor vehicles invariably have power applied to only two wheels, the limit of safety in speed depends largely upon the weight of the vehicle, always granting that perfect brakes are used; so that the practical or safe limit of speed of a motor vehicle depends upon its weight, as the amount of momentum stored up in a given weight on any vehicle cannot be brought under control by brake application beyond the point of locking the wheels, any more than that of a car can. In vehicles weighing three thousand pounds, the writer's experience has been that twelve miles an hour can be run with perfect safety, as the vehicle with rubber tires can be brought to a full stop in about eighteen to twenty feet. In light road buggies, weighing a thousand pounds, sixteen miles an hour are perfectly practicable and safe, as these buggies can be brought from this speed to a full stop in about the same distance, and so on for any intermediate weights; so that the safe speed on city streets for an electric vehicle weighing from one thousand to three thousand pounds is from twelve to sixteen miles per hour as a maximum. And when the day comes, as it will very shortly, that city ordinances will be passed governing the speeds at which motor vehicles will be allowed to operate, these conditions should be taken into consideration and the lighter vehicles allowed a somewhat higher speed than the larger and heavier ones. When speeds are considered in the passing of such an ordinance, three or four divisions, from sixteen or twelve miles down, should be considered sufficient, as at six miles an hour any electric carriage can be stopped in about its own length. No restrictions were placed on the speed of automobiles in France for a long time, but during the last year a very strong sentiment has arisen in the automobile clubs of that country against reckless driving, and it is to be hoped that it will be taken hold of in time in the United States to prevent the difficulties and distress that overtook many of the French owners of automobiles, which, however, were not in any general sense electric vehicles. Manufacturers of automobiles in the United States are confining themselves, without exception almost, to a maximum speed within the limits cited, and as very few accidents have occurred in the use of electric automobiles from fast driving, it is safe to say that when city councils do take up the question, all of these things will be considered, and a reasonably fast speed allowed automobile users. Surely twelve to sixteen miles an hour is fast enough when we consider that in city work it is practically double the speed and one-half the time made by the horse. On the first introduction of motor vehicles, a great deal of stress was laid on speed, articles appearing in many papers advocating extremely high speeds, and in some of the tests and races given in this country, nearly fifty per cent of the value of an automobile as a prize winner was placed in its speed. The absurdity of this, however, is already proven. Almost any speed desired can be obtained from an electric automobile. It is simply a question of power application and gearing, and those who wish to ride at breakneck speed should only be

34

allowed to do so in places provided for the use of high-speed and racing machines, as roadways or racing tracks set aside for their exclusive use, which will no doubt in time come in the same way that they have been provided for horses.

CHAPTER III

SECONDARY OR STORAGE BATTERIES—THEIR CONSTRUCTION, USE AND OPERATION IN AUTOMOBILES

Probably in the whole history of mechanical and electrical devices there has been nothing of more importance and at the same time so little understood by the general public as storage batteries. But as the problems connected with them, from this time on, must be placed in their hands to a greater or less extent by the introduction of electric automobiles, it is necessary that they be understood.

One buys a fine carriage and a fine harness, both of which are useless without the horse. He then purchases a horse, and every care and attention is lavished upon him for his comfort, maintenance and longevity. In the same sense, when one buys an electric carriage the storage battery part of it is its propelling power or horse, and should have every care and attention to ensure its perfect working and longevity that is given the horse. No matter how well a carriage may run, no matter how perfect the motors, controllers, carriage design and all of its appointment, they, in themselves, are valueless in operation without the storage battery, and as this constitutes the soul or moving power of the vehicle, no expense or attention within practical limits should be spared to preserve its usefulness and ensure its perfect operation.

I am asked by hundreds of people, "What is a storage battery? How is it made and how does it work?" and this from people who belong to the better-educated classes. It has never been necessary for them, up to the present time, to inquire into these things and they are always very frank in confessing their ignorance of the matter. Manufacturers of automobiles, therefore, do not expect that these people shall understand these things until they have been educated either by experience or proper and careful instruction, or both.

In an electrical dictionary, written by Houston, we find storage cells described thus: "Two relatively inert plates of metal or metallic compound immersed in an electrolyte incapable of acting considerably on them until after an electric current has been passed through the liquid from one plate to the other and has changed their chemical relations."

This explanation in itself is confusing to those not versed in scientific or electrical nomenclature. From this description, a storage battery in its simplest form would be made of two plates of lead, one called a positive plate and the other called a negative plate, on which plates special preparations are used or made, called peroxides or oxides, and are obtained in as many different ways and by as many different processes, almost, as there are different makes of storage batteries depending upon their mechanical and electrical construction.

Without going into all the processes of manufacture, which do not concern the user, we shall at once proceed to an understanding of how and why they operate in their functions of storing up electricity. The two plates, being placed in a containing cell, usually a rubber jar for automobile work, are then submerged in what is called an electrolyte, which consists of a mixture of commercially pure sulphuric acid and water, being about eight parts water to one of acid, the exact mixture of which will be given later on. They are now ready for charging. These two plates are separated entirely from one another and have projecting from the containing cell two lugs or terminals to which attachment is made from some outside source of electricity, the positive pole being attached to one plate and the negative pole to another plate. The electric current, entering by one plate into the cell, passes through the liquid solution to the other plate and from the other plate out to the external electric circuit again. Its action in doing this decomposes the electrolyte or liquid between the plates, depositing the electro-positive results of this action on the negative pole of the outside source of supply and the electro negative results of this action on the plate connected with the positive pole of the outside source of supply; the sum total of both actions being the formation of hydrated sulphate of lead. After the battery has been so charged and the external or outside source of electric current disconnected, it will stand practically inert until a continuous wire forming a closed circuit be connected to both the positive and negative poles outside of the solution. Then it produces a current which flows through the liquid from the plate that is covered with the electro-positive results to that covered with the electro-negative results, or in an opposite direction from the current which was used in charging it. The result is a decomposition of the hydrated

38

sulphate of lead which was formed in charging. This, in general, is the principle of operation of any and all storage batteries.

The facts are that nothing like direct storage is obtained in the sense in which the word is usually understood. A chemical change is simply created by the passage of the electric current in charging, which afterwards, being connected to a closed circuit of its own, causes the battery to reverse its action in trying to assume a normal condition and in so doing generates an electric current.

However, the construction of storage batteries is not confined to two plates in one cell. Theoretically, any number can be used. (See Plate VII, which shows a seven-plate cell.) Common practice has as high as fifteen plates contained in one jar. Automobile cells usually have three, five, seven, nine, eleven and thirteen, according to the capacity for storage that is desired. The size and shape of the plates is simply a convenience of assembling and capacity desired. Mechanical construction and support have to be considered in proportion to the size, weight and number of the plates.

The voltage of an electric circuit of any kind is its measurement of strength or pressure in the same ratio that pounds measure the pressure of steam or compression of water in hydraulics. The amperage or amperes of an electric current is a measure of its volume or quantity, in the same ratio that water is measured by gallons, and the result of the two multiplied together is termed watts, which indicates the total amount of power—all of which is without reference to any time unit. Thus 100 volts multiplied by 10 amperes equals 1,000 watts, commonly called a kilowatt, and its equivalent in horse-power is always determined by dividing any number of watts by 746, which is the number of watts contained in one horse-power.

This is all stated in a very simple form for the purpose of giving the reader a clear understanding as to the conditions and relations of units by which he makes his purchase of electricity for operating his automobile. The price for electricity varies in different places from five to ten cents per kilowatt and in others from five to ten cents per thousand watts. This is figured by the manufacturer of electric current in this way: Ten cents a thousand watts means ten cents for a thousand watts flowing one hour; or ten cents per kilowatt hour means ten cents for a thousand watts or one kilowatt flowing one hour.

To illustrate: If one were charging an electric vehicle on a 100-volt circuit (which would be the standard pressure of the circuit and would not vary in its measurement, as all electric circuits have a standard voltage which is kept constant), and twenty amperes were allowed to flow for one hour, 2,000 watts would have been consumed or two kilowatt hours or 2,000 watt hours, which, at five cents for either name used, would amount to ten cents' worth of electricity. The volume or quantity of electricity, as before stated, is its amperes and is variable to any extent that it is desired to make demand for, and the term employed in using amperes when time is indicated is ampere hours. In different sizes of storage batteries the capacity of the battery is rated in ampere hours, which is obtained by the number of amperes flowing, multiplied by the number of hours which they flow. Thus, if a storage battery were given ten amperes for ten hours, its capacity would be one hundred ampere hours, and if this same battery were to give fifty amperes for two hours it would also have a capacity of one hundred ampere hours, and so on for any other division between amperes and time at which it is desired to use the battery.

It is also necessary to explain here that ampere hours have nothing to do with kilowatt hours or watt hours, because no consideration of voltage is given in using such an expression. Voltage is sometimes called potential, and the technical expression in referring to a storage battery to get at its voltage is that between the two plates in its chemical changes, a difference of potential equal to two volts is obtained, and lead is used for storage battery because no other metal or material has ever been found that has such a big difference of potential combining with it the necessary elements of durability.

One cell of storage battery, no matter whether it has two or any other number of plates, never has a greater difference of potential than an average of two volts, so that a cell that would give one ampere for ten hours or ten amperes for ten hours or a hundred amperes for ten hours, would individually have but two volts. If we wish to charge from a 110-volt circuit, commonly used and obtainable, we must have at least forty of these cells connected up in series, which, at an average of two volts per cell, would make eighty volts, which would still give us thirty volts to spare; but, as in charging up to a maximum the voltage rises to about 2.2 volts, and as these cells have some internal resistance offered to the various connections and wires from the external source to the

41

batteries, usually forty cells of battery require from 106 to 108 volts, and are commonly charged from a 110-volt circuit. To figure out the complete process of charging and cost thereof would be as follows:

A 100-ampere hour battery charged from a 110-volt circuit at the rate of ten amperes would require ten hours to charge and would consume in that time from the external source of supply 110 volts, multiplied by ten amperes, or 1,100 watts per hour, which, multiplied by ten hours, would be 11,000 watts, which, at five cents per thousand watts, would cost 55 cents. The cost per mile for operating an automobile would then simply be figured from the number of miles that a 100-ampere hour battery will run a vehicle. Thus, if a vehicle will run forty miles on one charge of the batteries, it will cost 1.37 cents per mile to run it.

Having now given in a very brief way the principles involved in the working, measurement, and method for obtaining cost in the use of storage battery, we will proceed to matters more to the point as regards its care and maintenance.

Batteries for automobiles usually consist of forty or forty-four cells per vehicle, varying with the different manufacturers, and as a general thing are divided into four trays with either ten or eleven cells in each tray. This is done for two purposes: For convenience in handling (by subdividing the weight) in getting them in and out of the vehicle; and because of the combinations of connections desired in the batteries to obtain a controller operation for different speeds, this latter function being obtained in a different way by nearly every manufacturer—

different but essentially the same in results. The speed of an electric motor is always in direct proportion to the voltage of the current supplying it. So, if forty cells or eighty volts will run a vehicle twelve miles per hour, twenty cells or forty volts will run it about six miles per hour and ten cells or twenty volts about three miles per hour. The latter is when all four trays are in what is known as parallel, i. e., they are the same as though we had only ten very large cells in the vehicle. The next combination is to connect them up as though we had only twenty cells in the vehicle; and the third combination, by connecting them up so that all forty cells are represented in the voltage, which give us corresponding speeds of the vehicle as enumerated, and at any speed or any combination in running the vehicle all of the forty cells are discharging current at exactly the same rate per cell, because it would be a disastrous thing to have any arrangement of batteries whereby some cells work longer or discharge more rapidly than others. All the cells in an automobile must be charged and discharged exactly alike and in the same time. Right here is the best place to say that if any one or more cells in any one tray of the vehicle becomes imperfect or disabled, it should be immediately attended to, because one tray with nine cells only will be attacked in the various combinations of control by the trays having ten cells in them, much to the distress of the cells remaining in the tray containing the lesser number of perfect cells.

It should be further said here that to maintain the uniformity of action in a battery, which means everything to its life and durability, every possible care should be taken to keep the connections clean and perfect in whatever part of the circuit they may take place. The controller contact should be kept clean and in good shape and the connections of the batteries should be gone over frequently to see that none of them are loose, because if they are, they introduce a certain amount of resistance into that particular tray which makes it labor under conditions not favorable for its working in harmony with the others. It is hardly reasonable to expect the storage battery to take the place of the horse and receive no attention whatever. If, in total, we consider the small amount of care required on the part of the battery as compared with the enormous amount of care, attention and trouble which it is necessary to give to the horse which the battery displaces, the care of the battery should be considered more of a pleasure than a task, especially when the importance of such watchfulness is so pronounced.

If properly cared for and looked after from week to week, the personal attention and expense necessary to bestow upon an electric vehicle are very small as compared with one that is operated by animal power. The main difficulty is that purchasers and users of automobiles are apt to become too negligent of their batteries and vehicle for the simple reason that they do require so little attention, and thus they do not get even that little that is their due, in some instances users assuming that they will require no attention at all.

It is only reasonable for manufacturers to require their customers to make the moderate preparations necessary for the proper maintenance and care of vehicles. This requires but a very moderate expenditure, equaling but a very small amount as compared with the amount required to harbor and maintain a horse and carriage. No electric vehicle should be purchased without the addition of a volt meter, ampere meter, rheostat and switch, to use in connection with charging the batteries, if the owner wishes to obtain good results, good economy, and long life to his batteries and vehicle. (See Plate VIII.) The few following rules should be observed in connection with this:

Never start out with the vehicle for any length of run without being sure that the batteries are fully and properly charged.

Be sure that the electrolyte, i. e., the solution in the batteries, always covers the plates at all times and in all cells.

An imperative rule is, always charge the batteries promptly after using the vehicle. Nothing will depreciate mileage capacity so quickly as to leave batteries discharged for any length of time (even an hour is detrimental), except operating the batteries without the proper amount of solution in them. The solution should at all times extend above the plates. In the event of its having evaporated, it should be replenished at once and the batteries charged at a slow rate until the cells read up to their full voltage of from 2.2 to 2.6 per cell, according to what make of cell is used.

The solution that should be put in the batteries when they have been overhauled and need a new solution consists of commercially pure sulphuric acid and pure water. By pure water we mean distilled water that is free from minerals of any kind. A hydrometer should be used and the electrolyte, consisting of the water and acid mixed, should show a specific gravity of 1.250, or 29 degrees Baume, when cool. One caution to be given is that the acid should always be poured into the water— never otherwise. The mixture should never be poured into the batteries until it is cool, as considerable heat is generated in the first combination. Batteries are always shipped by manufacturers with the proper solution in them and every precaution is taken to have them reach the purchaser in a perfect condition; but occasionally accidents will happen in transportation, and it is always necessary, upon the receipt of a vehicle and batteries, to examine the latter and see that the electrolyte is well over the tops of the plates up to within one-fourth or three-eighths inch of the top of the rubber jar.

When it becomes necessary to add to the solution in the cells because of the evaporation of the water, a very weak solution should be used as compared with that put in for filling the cells with entirely new solution, which is about eight parts water to one part acid, while the solution for touching up or replenishing the cells should be ten or eleven parts water to one part acid. In charging batteries, great care must be taken to leave the doors to the openings of the carriage body open, as often during the process of charging a great deal of hydrogen gas is thrown off; and care must be taken that no spark, lighted match or flame of any kind is introduced near the box of the vehicle at such times.

In charging at high rates, great care must be taken not to heat the cells. If at any time cells show a tendency to get hot while charging, the charging current must be immediately reduced. If, under normal

46

conditions of charge, cells that heretofore have shown no evidence of heating should show heat, an inspection as to their condition should at once be made.

The writer has used, and is using, a great many Willard batteries, and as in charging at high rates, to get the best results, their tables and instructions have been found to be accurate, they are introduced here for the benefit of the general reader.

"CAUTION.—In charging at high rates, understand as near as possible the discharged condition of the battery, as the highest rate should only be employed when the battery is completely discharged."

The size numbers of the Willard cells correspond to the size numbers used by the Woods Motor Vehicle Company, as follows:

Willard	Woods	Willard	Woods
No. 1001	No. 1½	No. 1109	No. 3
No. 1005	No. 1	No. 1111	No. 4
No. 1007	No. 2	No. 1113	No. 5
		No. 1115	No. 6

AMPERE HOUR CAPACITY WHEN DISCHARGED IN 3, 4, 5 OR 6 HOURS

Cell No.	3 Hrs.	4 Hrs.	5 Hrs.	6 Hrs.
1001	34	38	40	42
1005	45	50	53	55
1007	66	73	78	81
1109	112	124	132	137
1111	140	155	165	171
1113	168	186	198	206
1115	196	217	231	240

TABLE FOR FULL CHARGING IN 45 MINUTES

Cell No.	20 Min.	5 Min.	5 Min.	10 Min.	5 Min.
1001	72 Am.	52 Am.	36 Am.	16 Am.	5 Am.
1005	96 "	68 "	48 "	20 "	7 "
1007	140 "	100 "	70 "	30 "	10 "
1109	238 "	170 "	119 "	51 "	17 "
1111	300 "	214 "	150 "	64 "	21 "
1113	356 "	254 "	178 "	76 "	26 "
1115	420 "	300 "	210 "	90 "	30 "

As an example: No. 1007 is charged at a rate of 140 amperes for the first twenty minutes; 100 amperes for the next five minutes; 70 amperes for the next five minutes; 30 amperes for the next ten minutes; and 10 amperes for the last five minutes.

Caution should be taken not to attempt to charge the battery at this high rate unless it be completely discharged, for it will be observed that the rate of charge that the battery will absorb is dependent upon the amount of energy already absorbed by the battery.

TABLE FOR FULL CHARGING IN 3 HOURS

Cell No.	½ Hr.	½ Hr.	½ Hr.	½ Hr.	1 Hr.
1001	36 Am.	20 Am.	16 Am.	10 Am.	5 Am.
1005	48 "	28 "	20 "	16 "	7 "
1007	70 "	40 "	30 "	20 "	10 "
1109	120 "	68 "	52 "	32 "	17 "
1111	150 "	86 "	62 "	42 "	21 "
1113	178 "	102 "	76 "	50 "	26 "
1115	208 "	118 "	90 "	60 "	30 "

As an example: No. 1007 should be charged at 70 amperes for the first half hour; 40 amperes for the next half hour; 30 amperes for the next half hour; 20 amperes for the next half hour; and 10 amperes for the last hour.

This rule must be carefully observed and the charge should not be continued beyond the time above given, nor after the battery reads 2.6 volts per cell at the last charging rate.

EIGHT HOUR CHARGING TABLE

Cell No.	Charging rate in Amps. for 8 hours	
1001	36	Amperes
1005	48	"
1007	70	"
1109	120	"
1111	150	"
1113	178	"
1115	208	"

This rate may be very convenient in most cases by allowing the battery to charge during the night, and this method is the most simple of any from the fact that it requires no attendance.

To be able to benefit by the whole range of possibilities in charging batteries as given in the foregoing tables, the circuit that supplies the batteries with current should be supplied with a volt meter, ampere meter and rheostat. In this way the voltage of the batteries can always be measured individually and collectively, or the amount of current it is desired to put into the batteries for any length of time can be measured by the ampere meter and the amount varied as desired by the use of the rheostat. All of the foregoing should be supplied by the automobile manufacturer at a reasonable charge in addition to the vehicle, as required for any of the vehicles he manufactures.

It is a very good idea, in fact necessary, in the care of batteries, to charge them at least once a month at a slow rate for a period of ten or twelve hours after they are fully charged. By a slow rate is meant at about half the ampere rate given in the eight hour charging table. This has a tendency to tone up the batteries and equalize any little difference that may exist in the cells. Thus, if a battery at any time should not read up to 2.2 or 2.6 volts per cell, according to the make of battery, it should be given the long charge, as indicated above, until the voltage is equal to 2.2 or 2.6 volts per cell, and if the voltage does not then come up to this, it is a clear indication that the cells are in an unhealthy condition from some internal cause. In this case, a reading should be taken of each cell individually with a low-reading volt meter, calibrated for this work, and as often as any of the cells in the set are found too low in their voltage, they should be disconnected by cutting the solid connecting bars in a "V" shape, so that they may be fitted together and soldered again, the cell taken from the tray, the element removed from the cell, the electrolyte discharged from the cell, and the cell washed out. If any sediment is found to have settled in the bottom from disintegration (which occurs to a slight extent in any battery), or between the plates, as it will sometimes lodge in the insulators between

50

the plates, this should all also be carefully washed from between them. They should then be put together again and new electrolyte supplied and great care taken in soldering and making the new joint after the cell is replaced in the tray.

If at any time a set of batteries will not retain their charge after being fully charged up, it is a very clear indication that they are short circuited. In some makes of batteries this is due to the sediment settling so rapidly in the bottom of the cell that it has come up to the bottom of the plates in the cell. Nothing will destroy a battery quicker than to try to operate it in this condition, and whenever a battery of any make shows indication of not holding its charge, or falling far below its normal capacity, it should be looked after in this respect without any loss of time.

Frequently difficulty is experienced in removing an element or lead plate from the rubber containing jar. In this case, immerse them in boiling water for a few moments and little difficulty will then be experienced in getting the element out of the cell.

A practical battery will run from 10,000 to 18,000 miles before it becomes necessary to renew the positive plates. When this does become necessary and the positive plates have been secured, the above method of overhauling the cell describes about the procedure that will have to take place in getting them separated, putting them back in place, etc. Although it is advisable to burn lead connections of the cells together instead of soldering them, when apparatus for doing so is at hand.

The instructions and information given here are as detailed as can be compiled to cover the general battery practice necessary to the successful running and care of an electric automobile. Each manufacturer will have some specific details on which information should be sought as regards his particular type or types of batteries, and the first effort of the purchaser should be to get all of the information and instructions that he can from whatever manufacturer he purchases his automobile, so that he may give it the same protection and care he would were he purchasing a valuable horse.

The cost of a charging outfit depends largely upon the size and capacity of the battery to be handled, but a fair estimate to make is from $25 to $75. Of course it must be understood that one charging outfit is capable of charging a vehicle several times a day or several vehicles a day.

The battery capacity for operating a vehicle, generally speaking, is now figured on a three-hour basis, i. e., if a vehicle of a given weight and size and for a given purpose requires twenty-five amperes to run it at its maximum speed, a battery is provided for the vehicle that practically gives twenty-five amperes for three consecutive hours. If a vehicle requires forty amperes, a battery is put in it that will give forty amperes for three hours in succession. Different sizes of vehicles and different constructions of vehicles require different sizes of batteries, but this is taken care of by the manufacturers of the different vehicles, and should be inquired into by the purchaser.

CHAPTER IV

ELECTRIC MOTORS, THEIR CONSTRUCTION, OPERATION AND CONTROL AS USED ON AUTOMOBILES

The electric motor is a device for transforming electrical energy into mechanical work, and having described the methods and means by which power is obtained for the propulsion of an electric vehicle, a not less interesting portion of an electric automobile is the means by which this power manifests itself in actual performance. To explain in terms void of any technical phraseology the operation and construction of an electric motor is almost impossible, nor will it be attempted here in minute detail, as all the science embodied in this wonderful piece of apparatus does not need to be made plain in a book written for the purpose for which this one is designed, namely, to give the purchaser of an electric automobile sufficient electrical information to enable him to select and care for it with discrimination.

Everyone knows that a magnet will attract the opposite pole of another magnet and will pull it around. Everyone knows that a magnet, or electro-magnet, will attract to itself anything in the shape of iron or steel. But probably everyone does not know that a magnet will attract a copper wire in the same way when an electric current is passing through such wire, and will not attract or be attracted by a copper wire when no electrical current is passing through it

Without going into all of the science pertaining to magnets, magnetic poles, lines of force, etc., we shall try to explain the operation of an electric motor on a more simple plane. After the invention of the electromagnet, which differs from the ordinary magnet by having a coil of wire wound around it at some one or more points, through which an electric current is conducted, whereas the simple magnet is merely a magnetized piece of steel or iron, many persons perceived that it would be possible to construct an engine in which one magnet should be pulled around by another, and that this could be varied in strength by the amount of current used in the two respective magnets; and it was also perceived that this rotation could be kept up continually by cutting off or reversing the current from one magnet to another as they pulled each other in succession around a circle. Out of this first understanding of magnetic power, the present perfected electric motor has grown, only instead of magnets being used altogether, powerful electro-magnets are built in the outside frame of the motor, around which large coils of wire are wound and through which an electric current is made to pass to energize them as magnets, and which correspond in this to the early conception and development of an electric magnet. Between these pole pieces is placed a piece of apparatus which is called its armature (see Plate X, showing armature resting on top of field frame) and which has taken the place of the magnets as first conceived to give a rotative effect. On this armature wires are wound in coils, varying in their number and in the number of turns to each coil according to the size and design of the motor. Each of the ends of these coils is connected to

54

what is known as commutator segments, there being a number of segments to correspond with the number of coils employed on the armature, the connection being diametrically opposite one another.

In some instances, where more than two magnets are used, these connections are made at segments one-quarter around the diameter of the commutator and armature.

When the armature is covered with wire the segments are all put in place, each segment of the commutator is insulated from every other segment; and, thus built together, they form a continuous and round surface on one end of the armature. To this commutator are applied what are known as brushes, situated either 90 or 160 degrees apart. To these brushes are affixed the wires that supply the electric current. The current passes through the brushes to one segment, then through the coil attached to that segment, around the armature by the several turns of wire of that coil, and the intermediate coil, and out at the other segment and brush; and as it has a certain angular displacement or distance on entering the coil away from the magnet toward which it is rotating, the magnet pulls the armature towards it in a rotative way by virtue of the attraction which the electricity in the coil has for the magnet. As soon as this particular coil has passed through under the brush, or face of the magnet, it ceases to be so actively energized and so passes by the face of the magnet and the current is switched back into the next segment, being at the same time reversed in the first one as it comes under the other brush, and so on for any number of segments or coils; and thus the rotative operation of an electric motor is kept up continuously; so that the real operation of an electric motor is the attraction of an electric current towards a magnet, the positive and negative polarities for this attraction being arranged by the proper commutation at the brushes.

While this is anything but a technical description of the operation of an electric motor, it will at least give those not versed in such matters at all a correct idea covering the principle of operation. These coils of wire are wound on frames giving them the right shape before being placed in the armature, and in case of repair can be obtained and easily inserted by any one familiar with motor repairing.

Now, the difficulties in an electric motor to be watched for are in this very commutation of the current from one coil to another in the armature. The brushes of electric motors for automobile work are invariably made of carbon. This has to be so because the adjustment and placing of the brushes has to permit of their being as permanent as possible and also to permit the motors to run backwards or forwards. They must have ample contact, and must be free from what is known as sparking—that is, a spark that is set up at the point of contact between the brushes and commutator where the current leaves the former and enters the latter, and which is usually brought about by an imperfect contact or an imperfect angular adjustment of the brushes; for the present art in motor design is one in which a motor properly adjusted and fitted with brushes will run under a great variation of load absolutely sparkless.

Whenever an automobile has been out on a run or is on a run and it is noted that the motors are getting warm or hot to any excess over their usual normal condition, in ninety-nine cases out of a hundred it is because the contact of the brushes with the commutator has become imperfect in some way, and nothing is more disastrous to a motor than to let it work under these conditions for any length of time. I have known motors to run for weeks or months without giving any trouble and then, because a brush had become too short or a brush-holder spring had loosened up, or some other simple thing of a like nature had been neglected, the motor would heat up frequently with damaging results, which, had the motors been regularly inspected once a week or once every two weeks, would never have occurred. The trouble is with motors, as with batteries, everything in their operation is concealed from sight, and when they run day after day and week after week without showing any visible distress, it is so very easy for the operator or owner to become negligent and think that they will run on in the same way forever without any attention.

No mechanical proposition has ever yet been conceived that has perfect infallibility, and it is very unfair to expect an automobile to be the first exception in this way. Proper attention to brush adjustment and to the keeping of the commutator clean and free from dirt and grit will prevent many interruptions to service, much expense, and a great deal of annoyance.

The electric motor for an automobile is a design and invention peculiar to itself. It can have no such protection or safety as is offered to motors in all other classes of work, like wires that will fuse when an excess of current is passing through them, or cut-outs that will operate for the same reason; because when such excess demand for power is made by the motor of an electric automobile, it is at a critical point in its operation, as on a heavy grade, a bad place to start, or a condition in which, of all times, power is wanted most and must be relied on.

Therefore, the motor must be designed and constructed to withstand all the power that can be applied to it in the propulsion of the vehicle for which it is built. It must stand the application of power by persons who do not use good judgment. I have seen men on delivery wagons, loaded heavily, turn on the last and strongest speed of the vehicle while it was standing still and fairly lift the thing off its feet—so to speak —all of which was entirely unnecessary and a very hard task for the motor. But motors are built now to stand this sort of thing, for weakness and excuses in their operation from such causes would not make them a practical proposition. But to stand all this abuse, and excessive work, they must have good care and the main point of attention is to keep the brushes in a perfect working condition. This cannot be emphasized too strongly. Any well designed motor is arranged so that it is accessible in this point in a moment's time, and it is only a matter of two or three minutes to inspect it and ascertain its condition, and only a matter of two or three minutes more to put in a new set of brushes if there is any question about their being in a worn-out condition. While every effort possible is made by every good manufacturer to make all connections of wires and contacts as solid and as permanent as possible, it must be remembered that a vehicle that is being bumped and shaken around on all kinds of streets is subject to all kinds of disturbances and these connections will, in time, wear loose. In some instances wire has been known to crystallize and break, but where proper flexible cables are used, the latter difficulty is a thing that rarely occurs; so that with this in mind, it is well to go over all of these connections carefully every week or two and see that they are solid and firm and well in place. This applies to the connections on the motors, the motor brushes, the motor poles, and to those leading from the batteries to the controller and on the batteries themselves, all of which should be arranged so as to be accessible for this purpose.

Motors more recently designed are supplied with ball bearings which are adjustable and interchangeable, and the adjustment of which should be occasionally looked after. (See Plate XI.) These motors run from 800 to 1,200 revolutions per minute, and the ball bearings need lubrication, which is usually effected by using what is commonly known as hard oil. Some apply it in one way, some in another. For the writer's own work, he is using a compression cup exclusively, and finds that one filling of this cup is sufficient to lubricate the ball bearings for two or three months, if once a week, at least, the compression cap be given a full turn, which in itself is hardly a second's work, but which, nevertheless, needs to be attended to.

And now we come to a question of considerable interest, and that is, the application of the motor to the vehicle and the work to be done. Some advocate a single motor equipment and some a double motor equipment. As for the writer, he uses both according to the work to be done. In a very light form of buggy, weighing one thousand to twelve hundred pounds, with a comparatively narrow wheel tread, a single motor equipment with a differential gear has been found to work most satisfactorily (see Plates XII and XIX), whether this was by means of a split axle direct or whether by the mounting of a differential gear in the motor through a hollow shaft, or with the differential gear on a counter-shaft, using a solid and fixed axle in the two latter cases; a differential gear must be supplied in this connection because, in turning comers or in turning the vehicle around, the outside wheel has to run faster than the inside wheel, and the difference in the speed at which they operate must be compensated for between the application of power to these wheels and the motor itself. The differential gear is introduced here for this purpose and this only. When the vehicle is running straight away, the differential gear is entirely inoperative.

On larger and heavier vehicles, the writer has met with the best success where he has used two motors, one connected to either rear wheel, in which each motor acts independently of the other so far as its power function is concerned. (See Plate XIII, showing one wheel and motor.) By a proper arrangement of control, these motors automatically adjust themselves to any difference of speed in the driving wheels as they turn comers or the vehicle is turned around; consequently no differential gear of any nature is demanded, one gear and one pinion being sufficient to form the connection between each motor and each separate rear wheel of the vehicle. In the earlier experiments the large gear wheel was either of brass or steel and the small pinion of the motor of either one or the other of these metals; but it was found in practice that they made so much noise that they were objectionable. Therefore, after a long series of experiments, it was found that rawhide could be successfully used in the small pinions, and that the use of rawhide practically eliminated all objections as to noise from the gears. The greatest difficulty with the rawhide pinion has been its short life, not from the work or power applied to it so much as from the constant wetting it receives when the vehicle is being washed, which changes its condition so much as to cause it rapidly to wear away. Usually a pair of rawhide pinions, when well taken care of, will give about two thousand miles before needing replacement.

The writer is now experimenting with an entirely new form of gear, with specially cut teeth in the large gear, being left open on the under side, and a roller tooth or a tooth with a roller bearing on it being made on the smaller gear or pinion, which also has no bottom in it between the teeth. This enables metal to be used on both gear and pinion, making them self-cleaning; they run absolutely noiselessly and have reduced the friction due to gears nearly fifty per cent. Up to the present writing, these gears have only been experimented with on the lighter class of pleasure vehicles, but it is almost safe to say that they will operate with equal success on the heavier types.

In connection with gears, it is only proper to remark here that they need lubrication as well as any other moving part of the vehicle and the lubrication best adapted to this purpose is a mixture of tallow and graphite, especially where rawhide is used, as this has a tendency from the tallow to make the rawhide impervious to moisture and at the same time acts with the graphite as a lubricant which is not easily displaced. Gears should be greased with a mixture somewhat similar to this every two or three days if the vehicle is in constant use, and the gears should be frequently cleaned with a stiff bristle brush. If they are kept clean and well lubricated it will add to their life many months of wear, will make them run noiselessly and increase the owner's comfort in the use of the vehicle.

The controller (see Plate XIV) for varying the speed of the motors, and consequently the speed of the vehicle, is arrived at by three different methods or combinations of wiring, varying somewhat in detail. (See Plates XV, XVI, XVII.) The first is what is known as series parallel control, which pertains only to the grouping of the battery cells in different ways by a proper switching device or controller and which supplies, as stated, but three different variations of the motor with three corresponding speeds. This, to illustrate, is as follows: Four rows of ten cells of battery in each row are placed in the vehicle, and a proper switching device or controller is provided whereby the four rows are first made to operate in parallel as ten cells only; second, two rows of two tens each are connected in series and these two rows in parallel, the result being the same as having two rows of twenty cells each; and, third, all four rows of ten cells each are connected in series acting as forty cells, the three combinations giving respectively twenty, forty and eighty volts.

The next method is the one adopted by the writer after having used the former for a long period of time, the difficulty with the former having been too great a difference of speed between the second and third combinations. If those combinations gave three, six and twelve miles per hour, the jump between six and twelve miles was found to be a little too harsh and, as six miles per hour was a little too slow, while twelve miles per hour was a little too fast in many cases, the motor fields were wound so that four combinations were obtainable. The first, being two rows of batteries in parallel of two rows each, giving forty volts, the first speed of twenty volts having been found practically useless on streets; the second being a paralleling of the fields of the motor, i. e., a combination wherein the windings of the motor field were reduced by one-half, increasing the speed of the motor about twenty per cent, and making a speed of about nine miles per hour, i. e., about half way between the six and twelve-mile combination; the third combination, being the batteries all in series, giving a twelve-mile speed in which the fields of the motor are again put back in series, the fourth combination being again to put the motor fields in parallel, giving nearly a fifteen-miles-an-hour speed on the level. But the plan of this last combination has been twofold in its purpose. In hill climbing, an extra amount of current is demanded over what is used on a level road, and as this increase of current makes an excessive number of ampere turns on the fields of the motor at a very considerable loss due to resistance of the wires of the fields, it was found that, by putting the fields in parallel, which amounts to reducing the number of turns of wire by one-half and increasing their cross section in proportion, much more effective work could be done on heavy grades and at a very marked increase in efficiency.

The third and last method used is by a series of combinations of the field windings of the motor alone, which can be made to operate at any speed and under almost any condition desired in this respect, but they have never proven nearly as efficient as the two former combinations and are very little used. Another and greater objection to this last method is that all the batteries being in series, if the circuit, or a lug, or anything should break in any one battery or in any one tray of batteries, the vehicle becomes inoperative, while with either of the two former methods this could not occur, as the vehicle would still be operative either by one, two or three rows of batteries.

A similar condition exists in the use of two motors in a vehicle. To illustrate: A vehicle that demands on a level road four horse-power for its maximum speed would be equipped with two two-and-a-half horse-power motors at their normal rating, or five horse-power in all. These motors, if properly constructed, should be capable of working from thirty to forty minutes at one hundred per cent overload, without any appreciable damage to them and without developing any excess heat. This would make a total power obtainable on demand from the motors of ten horse-power. Under these conditions, with either motor disabled from any cause whatever, it can be cut out and the vehicle operated with perfect safety on the other motor, as it is capable of developing five horse-power for quite a period of time. The operation of the brake, controller and reversing switch of the motors is reserved for treatment in the chapter on vehicle operation.

CHAPTER V

TESTING AND INSPECTION OF ELECTRIC AUTOMOBILES

A friend in need is a friend indeed, and it is to be hoped that the same general courtesy in times of trouble will be established among automobilists that finally came into existence among bicyclists; but the best friend that one can have in a time of need is his own self-reliance by a thorough and accurate knowledge of what to do when occasion requires, and in this chapter the writer will endeavor to set forth simple methods by which the owner of an electric automobile may himself discover and remedy little things that would otherwise appear to him to be very large things.

The first thing that should be ascertained upon the purchasing of an automobile is the amount of current that it consumes when on jacks, i. e., when jacked up in the bam and allowed to run free at first speed.

The next thing to be ascertained is the amount of current that it consumes when it is running at maximum speed on a hard, level street, which, having once been recorded, will always be an indicator in regard to the condition of the adjustment.

For instance, a light road buggy should run at its maximum speed on a hard, level street with fourteen amperes of current. If upon the use of an ampere meter it is found that the vehicle takes twenty or more amperes at third speed on a level street it is the clearest indication that something is wrong, for the vehicle is running hard and as a consequence taking more power than it should; but pains must be first taken to know that the ampere meter calibrated correctly.

When such a condition is found to exist, the wheels should be removed from the ball-bearing axles, inspected for broken balls, lubricated, and put back again. If nothing is found wrong here, the motors should be inspected as to their adjustment and should be carefully looked over and lubricated.

The next point of investigation will be the gears. The teeth themselves may be found to be injured in some way, or so dry that they are laboring with heavy frictional losses. If the gear teeth are found to be perfect, they should be well lubricated. After this the brushes of the motor and the tension of the springs on the brushes should be looked after, as often a great deal of unnecessary friction is found to exist at this point, and when the brush friction is very heavy, it not only makes a greater current consumption, but has a tendency to make the motors run hot from the excessive friction so created. The brushes need just sufficient tension to hold them firmly and well on the commutator and not so much as to make them grind the carbon up into powder.

If the commutator is found blackened and discolored, it is because the brushes have become out of adjustment and sparking, or excessive heating has taken place. This is the only point in an electric motor that needs constant attention and care.

If all the points enumerated have been carefully gone through, inspected and lubricated and put in proper adjustment, the vehicle will invariably be restored to its normal current consumption.

When operating a vehicle on the street, if it suddenly refuses to respond to any one of the speeds, the reason can invariably be attributed to a broken or loosened connection at some point. This may be a controller contact, which can be easily ascertained by an inspection of that. It may be a broken wire from the controller to the batteries, which also can be easily ascertained. It may be a lug on one of the batteries, which can be ascertained by measuring up the voltage, using a volt meter to see if current is flowing in every tray of batteries.

Difficulties may arise at times from either the controller or reversing switch contacts, through dust or grime or pieces of extraneous matter between their surfaces. Usually when a vehicle refuses to operate on third speed it is because a connection or contact has suffered somewhere and thus interrupts the circuit, but, with a controller series paralleling the batteries, it will operate at both the first and second speeds under these conditions; and with one using both field and battery, it will operate at the first and second speeds, but not the third and fourth. So, in troubles of this kind, the first point of attention is the contacts of the controller and reversing switch, and next the connections of the batteries in the trays themselves.

Sometimes the vehicle will refuse to operate at any speed, in which case first ascertain that the reversing switch is in proper condition, that the controller contacts are all right and that the vehicle will take current in charging. If it will take current in charging it is a sure indication that the difficulty is in the motor or motors themselves, and can be nothing more than a broken wire leading to the motors, or the bad contact of the brushes on the motors.

All of these troubles with bad contacts, broken wires, etc., in explanation seem like a multitude of sins, but the facts are that they are possibilities rather than probabilities in a well designed and constructed vehicle. Still, the best of families will have accidents, and this law holds good in electric automobiles. Because the vehicle runs well to-day, there should be no cause for surprise if something goes wrong tomorrow if these troubles are not anticipated and prevented by proper inspection from time to time.

Three points should be reiterated here, viz.: The solution in the batteries *must not* be allowed to get below the plates; the batteries must not be allowed to stand *discharged* even over night; and the motor brushes *must have* careful adjustment and *constant inspection* for adjustment.

If these three points are taken care of and the batteries charged fully, as per tables and instructions in chapter on batteries, the difficulties of operating an electric automobile are not ten per cent of what they are in driving a horse.

It must be again reiterated that when a battery cell or cells are found inactive, they must be immediately replaced with new ones in order that the other cells may be kept in operative equilibrium. Should one or two cells in a tray of ten become inactive and the tray be allowed to

69

remain in this condition, it is only a question of a very short time before all of the rest of the cells in that particular tray will be in the same condition.

In every city of any size in the United States, at the present time, there is an electric-light plant—which, of course, must be taken into consideration in the purchase of an electric carriage—and in this electric-light plant are employed electricians for its various branches who would be only too glad, for a small consideration, to inspect a vehicle once every week or two weeks at the most, and so anticipate and prevent troubles that might otherwise occur from negligence or ignorance regarding the matter. The writer is now drafting a schedule for the establishment of regular inspectors by whom every electric vehicle in the city of Chicago will be thoroughly inspected once every two weeks at a nominal charge of one dollar per inspection. By this small expenditure of two dollars per month, at least twenty dollars' worth of annoyance and dissatisfaction will certainly be saved. This proposition has met with the universal endorsement of all those who have or are about to purchase electric automobiles, and to persons in cities outside of Chicago the writer would most strongly recommend this same method of procedure.

Probably no query is more frequently made by the intending purchaser of an automobile than the one about the manner of ascertaining the condition of charge in his batteries when on the street, always under the supposition that he starts out with his batteries fully charged.

There is an instrument made by several electrical concerns called a volt-ammeter that costs twenty-five to fifty dollars, according to the size of the instrument, which performs a twofold purpose. The ampere meter is so arranged as to indicate the amount of current being taken by the vehicle at any time when it is running, and inversely also to indicate the amount of current that is being put into the batteries at any time when it is charging. The volt-meter, which is another scale on the same instrument, with another indicating hand or pointer, tells the voltage of the batteries at any given time when the vehicle is either charging or discharging.

When forty cells of batteries are fully charged, the volt-meter should register when the vehicle first starts out practically eighty-eight volts, or two and two-tenths volts per cell. The point to which batteries can be safely discharged is sixty volts, or one and five-tenths volts per cell, and as the batteries discharge, the volt-meter will slowly settle from eighty-eight volts down to sixty volts, and any intermediate point between sixty and eighty-eight volts shows the amount of current that has been used, or that is left in the batteries.

This is the only known method that has any accuracy at all for indicating the condition of the battery charge, and the instrument has been found very convenient, as it indicates both the amperage and voltage and is very accurate when first put into use; but the difficulty so far has been that it is almost too delicate to stand the bumping around that it gets on the streets, and consequently gets out of adjustment after being in use but a comparatively short time on the streets.

The speed of the vehicle varies also. If it runs fourteen miles per hour at eighty-eight volts as a maximum, and at sixty volts as a minimum, there will be just that ratio of difference in speed. After one has used an electric automobile for a few days, and has actually had the experience of running out of current, he can tell almost to a nicety the exact condition of his batteries by the way his vehicle is running. The writer has made inquiry of driver after driver concerning the use of volt-ammeters, or, as it is commonly termed, the drop of potential method of telling the condition of battery charge, and finds that they pay little or no attention to the instrument. Aside from this, nine men out of ten have specific territory over which they ride, and two or three trips over a given territory determine for a man exactly where he can go and what he can do with his automobile, and so a geographical indication becomes of as much importance to him as anything else, and the instruments are much more reliable when used only at place of charging.

The writer thinks that the best experience that a man can possibly have when he begins to use an electric automobile is to use it once or twice to its limit, and so familiarize himself with it, and thus know exactly how it works.

Again, many people ask, "What shall I do when I wish to go out into the country, or from town to town?"

The reply is that, if between towns the distance is greater than the vehicle's radius of action, it is wiser not to attempt it unless it is known where the batteries can be recharged. Twenty miles is the radius of action of any horse that is driven with reason; and whenever people go out to ride with a horse, the question of going out into the country and the question of coming back from the country and the question of the horse's endurance are things they never think of because they have become so accustomed to his limitations that they have ceased to comment upon them.

All batteries in a well-designed vehicle are arranged so that they can be charged in the vehicle, and as they are put up in trays, are also removable in a very few moments' time, so that duplicate sets can be substituted for discharged ones if it is necessary to have the vehicle in constant use; and this is certainly no more work than it is to change a horse, or pair of horses, on a wagon.

There is one point in connection with charging batteries in a vehicle that does not seem to be very generally appreciated and understood, and upon which warning is hereby given.

As batteries are charged in a vehicle they give off a sort of atomized spray, especially when they are charged rapidly, and as this spray contains a greater or less amount of sulphuric acid, which settles down over the outside of the cells and trays, and on the bottom of the wagon, something of a leakage of current is set up, as this mixture with acid in it is a great conductor of electricity.

While this condition is unavoidable and the leakages so set up are small, still in time they amount to a good deal. As a preventive of this, the trays themselves are painted with an asphaltum paint, as are all connections and terminals of the batteries. This does not prevent the leakage, but only protects the trays and wood work from being

73

impregnated with the acid. To check this as much as possible, when the batteries are taken out every two or three weeks to add a little solution to the cells, the whole inside of the wagon should be thoroughly washed out with water, the bottom being scrubbed with a brush, and the cells should also be washed down and scrubbed with a brush to remove as much as possible any of the solution that has vaporized itself over them during the process of charging.

Right here it is pertinent to speak of the washing of an electric vehicle. The motors of an electric vehicle are supposed to be waterproof, but the facts of the case are, it is almost impossible to make them so, especially where a high pressure stream from a hose is used. Therefore, in showering down a wagon, it is best to have a little oil-cloth to throw over the top of the motors, and so prevent any possibility of leakage to the inside, as water, once in, works no end of damage to the motor in its insulation and windings. Should it be discovered that water has gotten into the motor and soaked it to any considerable extent, the motor should immediately be taken out and given a thorough drying in some hot place, as over the top of a steam boiler, but where it cannot possibly be burned in the process.

Also keep water out of the gears that run the carriage, by keeping them thoroughly lubricated at all times, and, if exposed, keep dirt out of them by giving them a good brush-over every two or three days with a stiff broom brush that will take all the dirt out of the bottom of the teeth, after which always lubricate them again. Gears well cased and sealed up should be looked after at least once every month.

The electric gong of a vehicle should be kept in perfect working order at all times, for the safety of the occupants of the vehicle, as well as of those outside of it, depends a great deal on the use of this gong. It is an instrument of inestimable value when properly used, and a torturing nuisance when indiscriminately and maliciously employed.

In turning from one street to another it should always be touched lightly in advance as a warning to any who may be coming from the opposite direction but cannot be seen or cannot see the vehicle. In approaching any street crossing the bell should be tapped lightly that those in covered vehicles may have warning and act accordingly, and in passing a vehicle at such a place the bell should also be lightly tapped that the driver ahead may not swerve in front of the automobile. But all of this does not necessitate a prolonged ring. A light tap once or twice is all that is necessary, is not annoying and will not frighten horses. When suddenly stopping on the street for any purpose, the bell should always be sounded that others behind may have their attention attracted toward the vehicle and thus prevent a rear-end collision.

Automobiles make much faster headway on streets than horse-drawn vehicles do, and where streets are wide enough to permit, the writer recommends holding the automobile well in the middle of the road. The automobile driver has thus a commanding view of the entire road before him and can pass teams going in either direction with much more safety and with much less noise and annoyance in using his gong. It is much better than running close to the right-hand curb and continually trying to force teams out of the way and pass them, for the automobile owner must remember that the horse-drawn vehicle still has its rights on the streets to as great an extent as the automobile from a legal point of view, and by priority of use, a greater right from a moral point of view.

CHAPTER VI

STREET OPERATION AND CARE OF ELECTRIC AUTOMOBILES

How long will it take me to learn to run an automobile, and how am I to know whether it is working properly or not, and if not, what are the necessary remedies? are questions that are ever uppermost in the mind of the prospective purchaser, and are frequently asked by the merely curious inquirer. While this is rather a hard matter to explain in the absence of an automobile to exhibit and exemplify with, the writer will try to throw such enlightenment on the subject as will at least do away with the uncertainty regarding these things that seems generally to possess the minds of those interested in automobiles.

The natural hand for ninety-nine per cent of people to use in steering is the right hand, and it also seems to be perfectly natural to manipulate the controlling mechanism or brake—whatever the combinations may be in this respect—with the left hand; and as in this country we turn to the right and keep on the right-hand side of the road, the proper place for a driver to sit in practice would be on the left-hand side of the vehicle. In some unaccountable way the location of the driver of a horse-drawn vehicle became changed about in America from what it is in Europe, especially in England, where the driver sits on the right-hand side and always turns to the left; while in America the custom is for the driver to sit on the right-hand side and turn to the right. There is no question but that this latter method has been the cause of many locked wheels and broken vehicles in this country. In addition to this reason for putting the driver on the left-hand side, it places the lever of the controlling device on the outside of the seat of the vehicle, which, were the driver to sit on the right-hand side, would have to be placed in the center of the seat, a very inconvenient place, coming, as it does, between the two persons occupying the vehicle, and especially inconvenient when robes and such necessities are considered, as they would always interfere with operation, unless the controlling lever were placed on the outside of the seat, in which case it would have to be operated with the right hand, while the vehicle would have to be steered with the left hand, and this seems to be an entirely unnatural method and very hard to learn.

Therefore, in view of all these things, nearly all electric automobiles of the present time are made to operate from the left-hand side of the driver's seat. The connection between the steering hand of the vehicle and the position it will place the wheels in, is one that will turn the vehicle in the opposite direction from the side towards which the steering handle is pushed, as this has been found to be the most natural and spontaneous method of steering a vehicle.

The hand-wheel is an exception to this, but this method of steering has met with such general disfavor in this country that levers are used almost exclusively now by all manufacturers.

There has been a great deal of controversy among both manufacturers and users of automobiles regarding the number of manipulations in the operation of a vehicle that should be incorporated into one handle. Some have maintained that the steering, speed, controlling, braking and reversing devices should all be arranged in one handle or point of manipulation, leaving one hand entirely free. An effort has been made by some engineers to accomplish this, but the liability of doing the wrong thing at the wrong time has finally overcome this disposition, and leading manufacturers of the present time are leaving the steering mechanism entirely free from any other function, as the safety of the vehicle's transit and the reliability of its management should not be jeopardized by being incorporated with other duties or manipulations.

The steering wheels of an automobile have been suspended in many different ways, and many claims have been made for each different suspension. Without exception the steering wheels have a turning knuckle or joint immediately at the base of the hub. That is, each wheel is pivoted on the end of the axle. (See Plate XX.) The axle for the wheel extends out in the proper direction and a lever for attaching the steering rods extends either towards the rear or front at an angle of nearly 45 degrees, for turning the axle where it is pivoted. The angle of this steering is not exactly 45 degrees. It is pitched inwardly or outwardly from a 45-degree angle, according to whether it extends toward the rear or toward the front of the vehicle, so that the wheels will be turned at an unequal angle. The purpose of this is a natural necessity. In turning a vehicle around, the inside wheel travels on a smaller circle than the outside one, and the angle for which it is set for any given degree of a circle must be as much shorter than that of the outside wheel as there is difference between the two respective circles they will make in turning around. Otherwise they will turn hard, as one wheel would be obliged to slip and would have a tendency to wear the tire flat from such slipping. Again, this condition of angular difference has a tendency to balance or steady the vehicle when it is running straight away.

Whatever the devices are for steering a vehicle, they should embody these general conditions, no matter what the particular form of mechanism may be.

In steering an automobile, the driver has only to learn that he does not need to hold on to the steering handle with a tenacity that will tire him out, a loose but sure hold of the handle being all that is necessary. In turning a vehicle, the radius of action of the handle should be such that a turn of three or four inches in either direction from the center will be all that is necessary for the general gyrations required on the street, and the radius of action of the handle for turning the vehicle around in its shortest circle (which in a well-designed vehicle should be inside of fifteen feet diameter) should not be to exceed twelve inches either side of the center. With this arrangement, the driver will not continually annoy a guest occupying the same seat with him in his manipulations of the vehicle.

The first developments of a steering device were towards one which would run over any obstacle without swerving the wheels to either one side or the other, but this rigidity was found undesirable and wholly

80

unnecessary, and in operating a vehicle on a comparatively rough-street, it has been found that a little whipping of the handle causes no very great inconvenience and is much easier on the vehicle and in its riding. A street that would require such rigidity of steering gear as some people seem to imagine necessary would be a street wholly unfit for any form of vehicle, horse-drawn or otherwise, to travel upon.

The summary of all this is simply that the most practical steering handle is one that is free from all incumbrances and free in its radius of action.

There have been some suspensions of wheels made where the pivoting was at the center of the hub, which, in passing over heavy obstacles, were a little steadier than those pivoted at the butt of the hub, but the necessary constructional weakness in such design and method have practically caused them to be abandoned.

The next thing of importance to steering is the arrangement, adjustment and handling of the controller, whatever may be its form of construction. Three and four different speeds have been found all that is necessary for the successful manipulation of an electric vehicle on the street, as racing machines are not considered in this respect. The arrangement is such that the slow speed must always be applied first and so on up to the highest, and in coming back from the highest to the lowest, the intermediate speeds also have to be passed through. In steering an electric vehicle on the street, the controller should never be pushed forward at once beyond the second speed, on which the vehicle

81

should be allowed to gather what headway it will before the third speed is applied. The greatest strain that can be given a vehicle, both in its motors and batteries, is to jump it, when starting, at once on its third or fourth speed. This should never be done except in a very bad place where the vehicle will not start on its second speed.

If at any time, in starting a vehicle in a bad place, it will not start on its second speed, or on its first speed, the controller should not be allowed to remain in either of these notches a second longer than the discovery that the vehicle will not start. The controller should then be immediately pushed on to third speed, and should it not start there— which will never be the case in a well-designed vehicle—all speeds should be immediately cut off and an investigation made as to the cause of the vehicle's immobility. This rule is meant to cover bad roads, muddy places, bad holes in which the vehicle, may at times become placed, and does not apply to the starting of the vehicle when something in itself should, perhaps, be inoperative and the road conditions good.

Sometimes, when a vehicle gets in a bad place like this, if it is reversed and run backwards for not more than a foot or so, and then given a little start on third speed, it will pull out of the difficulty; but such conditions in roads should be avoided, as nine times out of ten when a driver gets into such a place in the road it is through his own carelessness or negligence in watching where he is driving the vehicle. When a vehicle refuses to start under normal conditions of road, it should be looked after as will be described later on.

Some vehicles have a brake that can be applied when current or power is applied to the motors, and great care should always be exercised that the brake and the power, by any combination, are never in force at the same time; and if such a brake is used it must always be ascertained that the brake is liberated after the vehicle has been standing before undertaking to start it up, and that the current is always cut entirely off from the motors, or is at the dead notch before the brake is applied. Otherwise, excessive and unnecessary strains will occur to the motors, batteries and gears.

However, a properly designed vehicle should have its brake and controlling mechanism arranged with some form of interlocking combinations in which the brake cannot be applied when the current is flowing to the motors, or, contrariwise, current or power cannot be given to the motors when the brake is in possession of the vehicle. Such an arrangement does away with all need of caution on the part of the driver from the foregoing conditions. On a level street the brake application is something that should be used discriminatingly. In starting up a vehicle the overcoming of its inertia requires from two to three times the amount of power that it takes to run it when under way at a given speed, and this inertia is stored up in the vehicle and given back when it becomes time to stop in momentum; i. e., the vehicle will run a long distance after the power is cut off. If well adjusted, with good ball bearings and constructional work, the vehicle should run five to six hundred feet on a hard, level road after the power has been cut off, from a speed of twelve or fourteen miles per hour. Now, if the vehicle is brought to a sudden stop by the application of the brake, this stored-up inertia is entirely thrown away, and if this occurs frequently during a run, a very noticeable mileage capacity will be dissipated that might otherwise be profitably utilized. When one desires to stop at a given point in some block, it is well to slow down from third to second speed and from second to coasting before the brake is applied for the final and full stop. By this method of procedure, especially on delivery wagons, which have to make very frequent stops, the mileage capacity of the vehicle will not be lessened much below what it would be on a straightaway run.

In making a bad street crossing, or over railway crossings, and such other bad places as will be frequently found in city streets, if the vehicle is running at high speed, it is far better to pull the controller back to zero and let it coast over, than it is to let the wagon run right on to the bad place before bringing it up to a slow speed by a severe application of the brake.

All these are simply cautions that an electric vehicle must have as much consideration and judgment used in the driving, as is given with a good horse, and if this be done many inconveniences, interruptions and annoyances will be saved the owner or driver.

All electric vehicles have at least two points for checking speed, and some have three. A favorite way of applying a brake is through what is termed a band brake, which is applied directly to the motor shaft, and thus transmitted through the same gears that are used for propelling the vehicle. In addition to this, some vehicles are fitted with a brake which is applied directly to the wheels, and is called an emergency brake. But this emergency brake has been found to be very unsatisfactory when applied to rubber tires and also very destructive. When the tires are wide, there is necessarily a great deal of friction, and when the brake is applied to the tires they will fleck and little pieces are apt to be broken out by reason of this friction and heat.

In addition to these two mechanical methods of applying the brake, a reversing of the motors can be used in case of an emergency, but this should never be done except when the controller is on first or second speed—never with the controller on maximum speed.

On some vehicles the reversing switch is so arranged in the controller that from a perpendicular position forward it runs the vehicle forward and from the other perpendicular it runs the vehicle backward. But this is a faulty design, as the reversing switch and controller should be separated completely.

To reverse the motors when the vehicle is under way is a very severe strain upon them, no matter what the combinations of controller and reversing switch may be, and it should never be resorted to except in cases of emergency, as when the other brake or brakes fail utterly to do their work.

While the foregoing rules are very general, they are absolutely reliable in principle; and no matter what make of electric vehicle is used, they are all points which should be carefully inquired into by the purchaser that he may give his vehicle proper and intelligent care and operation

CHAPTER VII

AUTOMOBILE CLUBS, MEETS AND RACES

Owing to the lack-of organization and the very limited number of meets and races that have been so far perfected and held in this country, the scope of this chapter necessarily becomes limited.

The best example of club organization is the "Automobile Club of America," with its principal office in New York City. The objects of this corporation are the formation of a social organization, or club, composed in whole or in part of persons owning self-propelled pleasure vehicles for personal or private use; to furnish a means of recording the experience of members and others using motor vehicles or automobiles; to promote original investigation in the mechanical development of motor carriages, by members and others; to arrange for pleasure runs and to encourage road contests of all kinds among owners of automobiles; to co-operate in securing rational legislation and rules for regulating the use of automobiles in city and country; to maintain the rights and privileges of all forms of self-propelled pleasure vehicles whenever and wherever such rights and privileges are menaced; to encourage the construction of good roads and the improvement of the public highways; and generally to maintain a social club devoted to the sport of automobilism throughout the country.

The Constitution of this club is as follows:

86

Article I—Name.

This corporation shall be known as "The Automobile Club of America."

Article II—Club Seal.

The seal shall be circular, with the words "The Automobile Club of America" inscribed thereon.

Article III—Objects.

Section 1.—The promotion of a social organization or club composed in whole or in part of persons owning self-propelled pleasure vehicles for personal or private use. To afford a means of recording the experience of members and others using motor vehicles or automobiles. To promote original investigation in the development of motor carriages. To co-operate in securing rational legislation and the formation of proper rules and regulations governing the use of Automobiles in city and country, and to protect the interest of owners and users of Automobiles against unjust or unreasonable legislation, and to maintain the lawful rights and privileges of owners or users of all forms of self-propelled pleasure vehicles whenever and wherever such rights and privileges are menaced. The encouragement and development in this country of the Automobile. To promote and encourage in all ways the construction and maintenance of good roads and the improvement of existing highways, and generally to maintain a social club devoted to automobilism.

Sec. 2.—The Automobile Club of America shall be essentially a member's club, supported by members' subscriptions, and not carried on for profit.

Article IV—Membership.

Section 1.—The membership shall comprise four classes, viz.:

a. Honorary Members.

b. Life Members.

c. Active Members.

d. Associate Members.

Sec. 2.—The honorary membership shall be limited to twenty-five, and shall include ex-officio the following: The President of the United States; the Governor of the State of New York; the Mayor of the City of New York; the Director of United States Road Inquiry.

Sec. 3.—The active membership shall be limited to four hundred, exclusive of life members.

Sec. 4.—The associate and life membership shall not be limited.

Article V—Government.

Section 1.—The officers of the club shall consist of a President, a First, Second and Third Vice-President, Secretary, Treasurer, and Consulting Engineer.

Sec. 2.—The general management and control of affairs, funds and property of the club shall be vested in a Board of Nine Trustees, exclusive of the President, First Vice-President and Secretary, who shall, ex-officio, be members of said Board.

Sec. 3.—All the officers of the club shall, at the time of their election, be the owners of Automobiles.

Article VI.

Section 1.—The Constitution may be amended only by a vote of two-thirds of all the active members present at a regular or special meeting called for that purpose.

Sec. 2.—No proposition to amend this constitution shall be acted upon at any meeting of the club, unless it shall have been presented in writing to the Secretary, signed by at least ten active members, and notice embodying the purport of the proposed amendment shall have been sent to each member of the club in the call for such meeting, which notice shall be sent at least eight days prior to the date of the proposed meeting.

The By-Laws are given here in full for the benefit of those who may wish to understand this club or have information toward the organization of a more local institution:

BY-LAWS

Chapter I—Government.

Section 1.—The general management and control of the affairs, funds and property of the Club shall be vested in a board of nine Directors, to be elected from its members as hereinafter prescribed and to be known as Governors.

Chapter II—Meetings.

Section 1.—The annual meetings of the club shall be held in the City of New York on the second Monday of October in each year. Notice of the time and place of holding the same shall be sent to each member at least ten days prior thereto.

Sec. 2.—Meetings other than the annual meeting may be called by the presiding officer of the Board of Governors, and a notice of the time and place of holding such meeting must be sent to each member at least five days prior thereto. The presiding officer of the Board of Governors must call a special meeting of the club when so requested in writing and signed by at least fifteen active members, and at such special meetings only the special business shall be considered for which the meeting was called, notice of the same being included in the call sent to members.

Sec. 3.—At all meetings of the club twenty-five active members shall constitute a quorum.

Sec. 4.—If a quorum shall not be present the presiding officer may adjourn the meeting to a day and hour fixed by him with the same effect as if held as above provided.

Sec. 5.—At all regular meetings the order of business, except when otherwise determined by a vote of those present, shall be:

1st. Reading and correction of minutes.

2d. Report of officers.

3d. Report of Committees.

4th. Elections.

5th. Unfinished business.

6th. General business.

Sec. 6.—Stated meetings of the Board of Governors shall be held on the first Monday of each month, at eight o'clock in the evening at the Club House in the City of New York. Special meetings of the Board of

Governors may be called by the presiding officer thereof, and shall be called by him on the written request of any three members of the Board.

Sec. 7.—Five members of the Board of Governors shall constitute a quorum.

Chapter III—Elections.

Section 1.—The officers and Governors shall be elected at the annual meeting of the club, to be held on the second Monday of October in each year. The election shall be by ballot. Each active member of the club in good standing not in arrears for annual dues shall be entitled to cast one vote for each office and a majority of the votes so cast shall be necessary to a choice. The polls shall remain open one hour and tellers shall be appointed by the presiding officer.

Sec. 2.—Nominations of officers, Governors, and for vacancies to be filled must be made by the Board of Governors as herein provided, and may also be made by any fifteen active members of the club in good standing, providing that such nominations shall be in writing and signed and received by the Board of Governors at least one week prior to the date of election.

Sec. 3.—The officers shall hold office until the adjournment of the next annual meeting or until their successors are elected.

Sec. 4.—At the annual meeting to be held on the second Monday of October, 1899, nine Governors shall be elected. As soon as possible after the election they shall divide themselves by ballot into three classes of three each. The term of office of the first class shall expire on the second Monday of October, 1900; the term of office of the second class shall expire on the second Monday of October, 1901; and the term of office of the third class shall expire on the second Monday of October, 1902, but all Governors shall hold office until their successors have been elected. At each subsequent annual election three Governors to serve three years shall be elected.

Chapter IV—Vacancies.

Section 1.—If a vacancy shall occur in any office or in the Board of Governors, such vacancy may be filled by a majority vote at a meeting of the Board of Governors, and the term of service of such officer or Governor shall expire at the time of adjournment of the next annual meeting; other vacancy, if any, thus created shall be filled by election at that annual meeting in the manner hereinbefore provided.

Chapter V—Duties and Powers of the Board of Governors.

Section 1.—At the first regular meeting after each annual election the Board of Governors shall appoint the following committees: (1) A Membership Committee, composed of three active members; (2) a House Committee, composed of five active members; (3) a Committee on Exhibitions, Contests, Runs and Tours, composed of three active members; (4) a Committee on Laws and Ordinances, composed of three active members; (5) an Auditing Committee, composed of three active members, who shall not be Governors nor hold any office in the club.

Sec. 2.—The Board of Governors shall:

a. Keep minutes of their proceedings and make a report to each annual meeting.

b. Nominate candidates for the positions to be filled by election at each annual meeting and cause such nominations to be sent to each member of the club at least two weeks prior to the date thereof.

c. Receive such other nominations as may be sent in to them in writing signed by at least fifteen active members in good standing, provided such nominations are received at least one week prior to the date of election, and cause notice of such nomination to be sent each active member of the club.

d. Cause a ballot to be printed containing the names of all the candidates nominated, which shall be distributed among the members at the time of the election, and no officer or Governor shall be balloted for unless nominated as herein provided.

e. Cause the annual report of the Treasurer duly audited as hereinafter provided to be promptly sent to each member of the club.

Sec. 3.—The Board of Governors shall have powers:

a. To appoint such additional committees as they may deem necessary.

b. To fix penalties for violation of rules or for conduct of any member detrimental to the welfare of the Club, and to enforce the same.

c. To remit penalties for offences against the rules and for accidental violation of the Constitution and By-Laws.

d. To elect members as hereinafter provided.

e. To make rules for their own government and for the government of the committees appointed by them, except as may herein be otherwise provided.

f. To perform such other duties as may devolve upon them in their official capacity.

Chapter VI—Duties of President and VicePresidents.

1. The President shall preside at all meetings of the club.

2. He may call special meetings of the club and he shall do so at the written and signed request of five members of the Board of Governors or of fifteen active members.

3. The First Vice-President shall assist the President in discharging his duties, and shall be chairman of the Board of Governors, and in his absence shall succeed to the functions and perform the duties which would devolve upon the President, if present.

4. In the absence of the President or First Vice-President, the Second or Third VicePresident shall act, and in the absence of all of said officers the Board of Governors shall elect one of their number, who shall succeed to the functions and perform the duties which would devolve upon said officers, if present.

Chapter VII—Duties of Secretary.

The Secretary shall notify each member elect of his election, and upon membership being perfected shall furnish him with the Club Book and report his name to the Treasurer, with the date of his election. He shall notify each member of each meeting of the Club, and shall give all such notices as shall be required. He shall keep a correct list of the members, with their addresses, numbered in the order of their election or reelection. He shall keep a record of all Club Runs, Tours, Contests and Exhibitions. He shall keep a correct roll of the Automobiles and the members of the club owning the same. He shall keep a true record of all proceedings at meetings of the club, with the names of members present, in a book provided for the purpose. He shall make an annual report to the club at the annual meeting and shall perform such other duties as may be enjoined upon him by the ByLaws or by the Board of Governors.

Chapter VIII—Duties of Treasurer.

Section 1.—The Treasurer shall receive all moneys of the club and deposit same in the name of the club in such bank or trust company as shall be approved by the Board of Governors and under the direction of the Governors shall disburse the funds of the club.

Sec. 2.—He shall keep regular accounts and submit same to the Board of Governors whenever so required. He shall so prepare and submit at the annual meeting a statement in writing showing the financial condition of the club.

Chapter IX—Duties of Consulting Engineer.

Section 1.—The duties of Consulting Engineer shall be to advise the club respecting Automobiles, public roads and highways and generally to act in an advisory capacity to the club.

Chapter X—Committees.

Section 1.—It shall be the duty of the Committee on Membership to investigate the character and standing of all persons proposed for membership in the club and to report upon the same to the Board of Governors.

Sec. 2.—The House Committee shall have the general management and supervision of the Club House, the Club Automobiles, and all employees of the club, subject to the direction, however, of the Board of Governors. They shall establish and maintain suitable House Rules. They shall regulate prices, order purchases, audit House accounts, receive and redress complaints and perform such other duties as may be assigned them by the Board of Governors. The amount of club indebtedness which may be incurred by the House Committee shall in no case exceed the amount appropriated to its use by resolution of the Board of Governors.

Sec. 3.—It shall be the duty of the Committee on Exhibitions, Contests, Runs and Tours, subject to the approval of the Governors, to make arrangements for, act as judges in, and take charge of all; Exhibitions, Contests, Runs and Tours, and to determine all routes, distances and conditions for Runs, Tours and Contests. Any appropriation made by the Board of Governors for the purchase of prizes for Exhibitions or Contests shall not be exceeded without the further consent of the Board of Governors, and all bills incurred in connection with Exhibitions or Contests shall be approved by the Chairman of the Committee prior to being paid by the Treasurer.

Sec. 4.—The Committee on Laws and Ordinances shall examine and report on all laws and ordinances in existence or that may be pending affecting the rights and privileges of owners and users of Automobiles, and shall take such action, subject to the approval of the Board of

Governors, as shall effectively maintain those rights and privileges whenever and wherever such rights and privileges are menaced.

Sec. 5.—The Auditing Committee shall audit the accounts and the annual statement of the Treasurer and shall report same to the Board of Governors at their regular meeting next preceding the annual meeting of the club.

Chapter XI—Membership.

Section 1.—Membership in the Automobile Club of America shall be of four classes, viz.:

a. Honorary Members.

b. Life Members.

c. Active Members.

d. Associate Members.

Sec. 2.—The honorary membership shall be limited to twenty-five, and shall include, ex officio, the following: The President of the United States; the Governor of the State of New York; the Mayor of the City of New York; the Director of the United States Road Inquiry. The Board of Governors may elect as honorary members any persons distinguished for their political, scientific, literary, industrial or administrative capacities. Honorary members shall be exempt from all dues, fees or subscriptions and shall have no right to vote at any meeting of the club, nor shall they have any right, title or interest in the property or assets of the club.

Sec. 3.—Life membership is secured by an active member commuting all of his subsequent annual dues and future assessments, by the payment at one time of five hundred dollars, upon making which he shall become a life member of the club; but in all other respects he shall continue to have the rights of an active member. The life membership is unlimited.

Sec. 4.—The active membership shall be limited to four hundred, who with the life members shall be constituted the owners of the club properties and the original managers of the affairs of the club.

Sec. 5.—Associate members shall be entitled to all the privileges of the club under such restrictions as are hereinafter set forth, and shall conform to its Constitution, By-Laws, and Rules, but shall not have the right to hold office or vote, nor any right, title or interest whatsoever in the property or assets of the club. Only persons whose place of residence or occupation is distant more than fifty miles from the City Hall, New York, will be eligible for associate membership. The associate membership is unlimited.

Sec. 6.—Each candidate for active membership must be proposed and seconded in writing and must be personally known to the proposer and seconder, both of whom shall be members in good standing and not members of the Board of Directors.

The proposal must state the name, address and occupation of the person proposed, and shall be sent to the Secretary of the club and delivered by him to the Committee on Membership. The name, address and occupation of the person so proposed, with the names of his proposer and seconder, must be sent to each member of the club at least one week before action is taken on such proposal, and shall also be posted in a conspicuous place in the principal room in the club. After due investigation and unanimous approval by the Committee on Membership the Board of Trustees may proceed to take action on the proposal. Elections by the Board of Directors shall be by ballot and two negative votes shall prevent an election. Notice of the election of a candidate shall be posted on the Club Bulletin, and the Secretary shall send him a notice of his election. If, prior to the election of any person proposed for membership, five members shall file with the Committee on Membership a written objection to such proposal the same shall be deemed to be withdrawn and shall not be presented to the Board of Directors for action.

All proceedings of the Board of Directors or Committee on Membership touching the election of any person proposed for membership shall be strictly confidential, and no Director or member of such Committee shall be questioned thereto.

Sec. 7.—No person shall have the privilege of the club until he shall have paid to the Treasurer the fees and dues, and if the same are not paid within thirty days after notice of election, the election shall be void.

Sec. 8.—Any member in good standing not in arrears or indebted to the club may resign his membership by delivering a notice thereof to the Secretary, who shall report the same at the next meeting of the club, and upon so resigning forfeits all his rights and interests in its property.

Sec. 9.—Candidates for associate membership shall be proposed and elected in the same manner as hereinbefore provided for active membership.

Sec. 10.—The Board of Governors may appoint to associate membership members of leading foreign Automobile Clubs (duly vouched for by the President or Secretary of such clubs) during their sojourn in this country for a period not exceeding three months without the payment of annual dues or fees.

Chapter XII—Entrance Fees and Dues.

Section 1.—The entrance fees and dues shall be as follows: For active members, fifty dollars, payable semi-annually, one-half on the first day of November and one-half on the first day of May of each year. For associate members, fifty dollars entrance fee and twenty-five dollars annual dues, payable semi-annually, one-half on the first day of November and one-half on the first day of May of each year.

Sec. 2.—The first one hundred members or those who make application on or before the first annual meeting and are accepted and elected as active members shall be exempt from the payment of the entrance or initiation fee, and are hereby styled "Founder Members."

Sec. 3.—An associate member may be elected an active member at any time by paying the amount of the difference in the entrance fee.

Sec. 4.—Any member whose yearly dues shall remain unpaid for ninety days after they shall have become due shall be deprived of all the privileges of the club until such dues shall be paid. The Treasurer shall promptly notify each delinquent member of the penalty incurred under this section, and shall also furnish the names of such members to the Secretary. If any member of the club shall be deprived of its privileges as aforesaid for two consecutive months, such member shall be deemed to have tendered his resignation and shall cease to be a member of the Club. A member thus terminating his membership shall not again be eligible until he shall have paid the amount so due, and may only be restored to membership by vote of the Board of Governors if a vacancy exists.

Sec. 5.—By a vote of three-fourths of the active members present at a meeting the club may levy one or more assessments upon each member not exceeding ten dollars in any one year, provided that notice of such proposed action shall have been given in the call for the meeting.

Sec. 6.—Any member who is absent from the United States for an entire calendar year shall be exempt from dues for such year, provided he gives notice of his absence to the Secretary prior to the first day of April of such year.

Chapter XIII—Club Badge.

Section 1.—The Board of Governors may adopt and provide such distinguishing badge or emblem for club members and owners of Automobiles as may be hereafter determined upon.

Chapter XIV—Indebtedness to the Club.

Section 1.—On the first day of each month or as soon thereafter as may be practicable notice of the amount of indebtedness of each member for supplies in arrears on the last day of the preceding month shall be sent to each member. If not paid on or before the 15th day of the month a second notice shall then be sent to such member, and if, one week after such notification, the indebtedness is not discharged the name of such member, together with the amount due, shall be posted on the Bulletin Board of the Club, and such member shall be reported to the Board of Trustees and shall be refused further credit until the indebtedness is discharged. No member's indebtedness to the club shall exceed the sum of one hundred dollars, and when such indebtedness shall reach the sum of fifty dollars notice shall be sent to such member and further credit and supplies shall be stopped until the entire amount due shall have been paid. If not paid within one week the member's name, together with the amount due, shall be posted on the Bulletin Board and also reported to the Board of Governors. The Board of Governors may suspend any member whose name has been reported for nonpayment of indebtedness for supplies, and unless such default shall be satisfactorily accounted for the Board may declare him to be no longer a member of the club, and his membership shall thereupon cease.

Chapter XV—Discipline.

Section 1.—A member may be admonished or suspended for conduct injurious to the welfare or character of the club by a two-thirds vote of the Directors at a meeting at which a quorum shall be present, but only upon written complaint of one or more members of the club or Board of Governors, and the Governors shall investigate the circumstances connected with such alleged misconduct and shall notify such member in writing of the charges against him at least eight days prior to said meeting, and an opportunity to be heard shall be given him at such meeting. Annual dues during the period of suspension shall be payable.

Sec. 2.—The Board of Governors may also in the same manner expel any member of the club for conduct which shall be deemed to warrant this penalty, provided he shall have at least one month's notice in writing of the charges preferred against him.

Chapter XVI—House Rules.

Section 1.—The Club House shall be open during the periods determined by the Board of Governors.

Sec. 2.—A member may personally introduce strangers to the club house for one day, recording their names with his own in the Visitor's Book. Strangers may, upon application of an active member to and with the consent of the House Committee, be admitted to the use of the club for two weeks. Residents of foreign countries, upon application of two active members, on the approval of the Board of Governors, may be admitted to the use of the club for six months on payment in advance of the annual dues. Members introducing strangers to the club will be responsible for their conduct while in the house, and for any debts that may accrue on their account

Sec. 3.—No member or visitor shall be allowed to give, under any pretense whatsoever, money or gratuity to anyone in the employ of the club.

Sec. 4.—No person shall take from the Club House any article belonging to the club, or from the reading-room, or library, any book, pamphlet or newspaper, or mutilate or deface the same. No subscription

103

or petition shall be circulated nor any article exposed for sale in the Club House without the permission of the House Committee. No game whatever for wager or money will be allowed in the Club House.

Sec. 5.—The club's Automobiles may be hired by members or seats may be engaged therein for Runs and Tours at the discretion of the House Committee.

Chapter XVII—Notices.

Section 1.—All notices required to be sent to any member shall be sent by mail prepaid to such member's residence or place of business, and such mailing shall be presumptive evidence of the service thereof. Any change in the address must be promptly sent to the Secretary, who shall report the same to the Treasurer.

Chapter XVIII—Amendments.

These By-Laws may be amended only by a two-thirds vote of all the active members present at a regular or special meeting of the Club, the purport of the proposed amendment having been stated in the call for the meeting.

Racing in this country has as yet not been developed in a marked degree, for the simple reason that the motor vehicles in this country are made for a different and more particular purpose, i. e., that of a carriage to ride in rather than a machine to race with. In addition to this the country roads of America are not adapted to such races as are given in

France, where they have roads as good as our city streets extending for hundreds of miles across the country.

The first attempt at racing was given under the auspices of the "Times-Herald" of Chicago, Thanksgiving day, 1895. Many entries were made for the race, but very few machines made the test, as the weather was unfavorable and some eight or ten inches of snow made the work done on that day very extraordinary, considering the undeveloped state of motor vehicles in this country at that period. They covered about fifty miles over the streets of Chicago.

This was followed by what was known as the "Cosmopolitan Race," given under the auspices of the Cosmopolitan Magazine, May 30th, 1896, in New York City, extending from the City Hall in New York to Irvington-on-the-Hudson, and return, the distance covered being fifty-two miles. The conditions and rules regulating this race were made in New York City and have been subject to much criticism.

From September 7th, to the 11th, of the same year, a race was given in Providence, at Narragansett Park, in connection with the annual Rhode Island State Fair. Twelve entries were made, eight of which appeared in the race. The rules covering this race were somewhat better than those that had been made before. The conditions were local enough to give all the carriages equal chances. In this particular series of contests the electric vehicle won the day. A very satisfactory report of this race is in the September, 1896, number of the Horseless Age.

A proposition of far more particular value than any of those mentioned was given in October, 1898, as a Motor Carriage Exhibition, at the "Mechanics Fair" in Boston, Mass. The rules and regulations covering these tests were of a more practical nature and the awarding of points for motor carriages had a much more practical consideration. In this, speed was not the main question. Control, appearance, general operation and utility of wagons, were given more consideration than speed. The coming year promises to draw forth a great struggle of all kinds in the automobile world. The writer would suggest here that these races be termed "Meets," except where racing is the distinct and only feature to be considered, and that a separate distinction be made between electric, steam and gasoline vehicles, as to what constitutes the practicability of the motor vehicle. Gasoline vehicles cannot, in some ways, accomplish what the electric vehicle can; on the other hand, the electric vehicle cannot accomplish what the gasoline vehicles do. This should have serious consideration as a proposition in allowing points on comparison. In racing pleasure carriages, points should not be allowed on vehicles that exceed a speed of twelve to sixteen miles an hour, which is simply a question of gearing and power application. The points allowed on speed above this should be confined exclusively to machines built for racing purposes only. Mechanically, much consideration should be given to points allowed on control and safety in operation of vehicle. The production of the vehicle as a carriage in finish, style, etc., should then be considered. The next allowance of points should be accessibility for inspection and repairs and then the general mechanical construction of details.

In the operation, serious consideration should be given in handling a motor vehicle up and down grades, not only as to the percentage of grade it will climb, but as to the safety of the vehicle and its occupants, in case of accident or emergency under such conditions.

The rules of the "Automobile Club of France" are very satisfactory in many points, and are reproduced here for the reference and study of those who have local interest in the organization of such clubs, meets, and races.

THE FRENCH RACING RULES

Preamble.

The ideas which have governed the drawing up of the present racing rules are the following:

1. The Automobile Club of Prance is the sole authority regulating races of Automobile vehicles and motor cycles.

2. The general spirit of these rules is that the races are run and won by a combination of the machine and its riders, which must not be separated during the race.

I.—General Rules.

Article 1.—Every competitor entering for a race of motor vehicles or motor cycles is supposed to be acquainted with these rules, and undertakes to abide, without dispute, by the results to which such rules may lead.

General Provisions.

Art. 2.—All automobile races and record trials organized in France shall be controlled by the Racing Rules of the Automobile Club of France.

Art. 3.—All races which are not controlled by these rules are forbidden, and all competitors therein will be disqualified.

Publication of Programme.

Art 4.—The programmes of races (1) must be sent to the Sporting Committee of the Automobile Club of France; (2) and must be published in the Press at least five days before the races, if they be on the track, or fifteen days if they be on the road.

Art 5.—The programme shall contain (1) the number of prizes and the amounts for each race; (2) the distances; (3) the amount, if any, of entrance fee attached to each event; (4) the date and hour for closing of entries; (5) the amount of forfeit, if there be any; (6) the place at which entries are received; (7) complete and exact itinerary of road races. These itineraries shall not undergo any modifications, except from absolute necessity; in such cases notice shall be immediately given individually to each competitor.

Art. 6.—After publication of the programme no modification shall be made in it as regards prizes—the amount of which shall not be increased—or as regards the nature of the races originally announced. Mention shall be made on the first page of all race programmes that the meeting is held under the rules of the Automobile Club of France.

Art. 7.—A copy of the programme and rules shall be sent to each competitor on his entering for a road race.

Classification of Races.

Art. 8.—Races shall be "open" or "reserved." "Reserved" races shall be races confined to competitors fulfilling a definition stipulated by the promoters.

Categories.

Art. 9.—The categories officially recognized by the Automobile Club of France are as follows: (1) Vehicles (motor cycles and small carriages) weighing under 250 kilogs. (5 cwt.); (2) vehicles weighing more than 250 kilogs. and carrying at least two passengers side by side, of an average of not less than 70 kilogs. (11 stone) each, it being understood that if the average weight of the passengers does not amount to 70 kilogs. (11 stone) each, the deficiency may be made up with ballast. In track races and records, however, vehicles with two seats need only carry one passenger, but in road races two passengers are compulsory.

In addition, the promoters may subdivide the two foregoing categories into as many classes as they please.

Art 10.—The Sporting Committee shall be the sole judge of the classification of all motor vehicles, as well as of questions which may arise therefrom.

Entrance Fee and Forfeits.

Art. II.—The amount of the entrance fee shall be fixed by the promoters, who will decide whether it is repayable or not to the competitors who have started.

Art. 12.—The forfeit is not a matter of right; it must be specified on the programme, as also its amount.

109

Art. 13.—Entrance fees which are repayable and forfeits, if they are not claimed within a month, shall become the property of the promoters.

Entries.

Art. 14.—Entries shall be made as follows: (1) By letter; (2) by telegram, confirmed by letter of the same date.

Art. 15.—Any entry which is not accompanied by the fee, or which is sent in too late, will be annulled ipso facto.

Art. 16.—Any competitor wilfully sending in a false statement will be prevented from starting, and will be liable to a fine.

Racing Names.

Art. 17.—Any competitor may use a racing name, subject to its being approved of by the Sporting Committee.

Art. 18.—The racing name becomes permanent and cannot be changed without the permission of the Sporting Committee, to whom a written request must be sent, accompanied by a fee of 20 francs.

Stewards of the Course.

Art. 19.—In every race upon the road or on the track the promoters shall choose three stewards, whose appointment must be approved of by the Sporting Committee, and whose names should be communicated at the same time as the programme.

Art. 20.—The stewards are entrusted with the carrying out of the programme, and with seeing that the rules are strictly observed, and are also to settle any protest that may arise out of the race.

Art. 21.—The stewards can either prevent a competitor from starting, or start him after the others, if his inexperience or the construction of his car would seem to present a danger to other competitors.

Art. 22.—The stewards have a right (1) to prevent a competitor from starting; (2) to publicly reprimand a competitor; (3) to impose fines up to a maximum of 200 francs (£8); (4) to disqualify a competitor for a maximum period of a month.

In these two latter cases the competitor has a right to appeal to the Sporting Committee.

Art. 23.—Should the stewards deem that a heavier fine ought to be imposed, they can apply to the Sporting Committee, which has full power to inflict any penalty after taking evidence from those interested.

Art. 24.—The starter is appointed by the stewards, and he alone can judge of the validity of a start.

Art. 25.—As a general rule the start is given while the vehicles are at a standstill, and they must start by their own means, but in certain cases a flying start will be allowed with the sanction of the stewards.

Art. 26.—The start shall take place in the order of entry, unless by special arrangement.

Art. 27.—In races upon the track the start shall be given to all the competitors at the same time, and this can also be done on the road, or the vehicles may be sent off with regular intervals between the competitors.

Judge at Winning Post.

Art. 28.—At the winning post there shall be one judge, and his decisions shall be final. If, however, there is a large number of competitors, the judge is entitled to assistance, but the judge must be chosen by the stewards.

Art. 29.—The winning of a race is judged from the front of the front wheel for motor cycles and motor carriages alike.

Art 30.—Should two or more competitors finish level the judge declares a dead heat, and the two prizes shall be equally divided between the competitors finishing level.

Art 31.—In distance races the competitor must cover the whole course in order to be entitled to a prize.

Art. 32.—In time races the competitors shall be placed according to the number of kilometers covered.

Art. 33.—When a single competitor starts a limit of time may be fixed by the stewards within which the course must be covered.

Art. 34.—Should a single competitor start in a race he shall have the right to the first prize.

Observers at Corners.

Art. 35.—Observers chosen by the stewards shall be placed at the corners of the course to see that one competitor does not interfere wilfully or otherwise with another by wrongfully getting in front of him, or shutting him in, or by any other maneuver which would be calculated wrongfully to affect the result of the race.

Observers in Road Races.

Art. 36.—In road races a certain number of observers shall be appointed and placed where it may be necessary to stop the competitors, or compel them to drive at a stipulated speed, and the observers shall see that these instructions are strictly adhered to by the competitors.

Track.

Art. 37.—The measurement of the track shall be taken at 0.30 metre from the inside ropes. On all tracks the winning post must be indicated by a clearly-indicated line.

Art. 38.—For the establishment of records on the track a certificate of measurement, with an annexed plan prepared by a qualified surveyor, shall be furnished.

General Regulations Relating to Races.

Art. 39.—Any competitor who in a race crosses in front of another, shuts in or obstructs another by any means so that the latter is prevented from advancing, may be stopped in the race or penalized by fine or disqualification, so long as the collision was not rendered unavoidable by a third competitor or the competitor who was obstructed was not himself in fault, but the fact that the collision was involuntary, or that it did not affect the result of the race, shall in no case be admitted as a valid excuse.

Art. 40.—No competitor shall be allowed to cross the course of another until he is at least two lengths of the motor cycle or motor car ahead of such other competitor.

Art. 41.—No sign or advertisement shall be displayed on any vehicle while racing.

Art. 42.—No vehicle shall be pushed or assisted by any one other than its authorized occupants, under pain of disqualification.

113

Art. 43.—Competitors shall be responsible for all civil and criminal penalties whatsoever.

Special Regulations for Track Racing.

Art. 44.—A competitor wishing to pass another must do so on the outside, and so as to leave the competitor passed the following space from the rope, viz., for motor cycles, 1.30 metres (4 feet), and for motor cars 3 metres (10 feet).

Art. 45.—A race containing too many entries may be run in heats, semi-final and final.

Art. 46.—The racing stewards shall arrange the heats, semi-final and final, and their decision shall be without appeal.

Art. 47.—No accident shall admit of a competitor running again, either in another heat or in the final.

Art. 48.—Any competitor leaving the track to get off his machine must start again from the point where he left the track.

Special Regulations for Road Racing.

Art. 49.—In road races the approach of a competing vehicle must be notified by a horn, trumpet, or some similar instrument.

Art. 50.—Vehicles which have to travel by night must carry and display a white and green light in front and a red light behind.

Art. 51.—In road racing, competitors must conform to the traffic regulations of the police.

Art. 52.—Competitors must make themselves acquainted with the route, and no allowance will be made for mistakes they may make. Moreover, if any competitor takes a shorter or easier route than the one prescribed, he will be disqualified. The stewards shall be sole judges of the comparative distance or ease of the routes followed.

Protests.

Art. 53.—The right of protest lies with the competitor, but the steward can always interfere officially in case of necessity.

Art. 54.—Any competitor lodging a protest must always substantiate his grounds of protest, and the competitor protested against has the right of being heard in opposition to the protest.

Art. 55.—No protest will be considered unless it is put into writing. Protests must be considered by the stewards on the spot, and a decision shall be reached immediately, whenever this is possible.

Art. 56.—Protests shall be lodged at the times and in manner following: Protests as to classification of competitors and of machines, as to validity of entry and payment of entrance fees—before the race and verbally. Protests as to unfair running, errors of route, or any other irregularities on the route—within twenty-four hours after the race, and in writing. Protests as to the fraudulent starting of a competitor in a race for which he was not qualified—eight days after the race, and in writing. For protests in races on the road—eight days after the finish of the race.

Penalties.

Art. 57.—Penalties imposed on competitors in and organizers of races are recoverable immediately upon notification of the parties concerned and upon publication in the journals officially notified by the Sporting Committee.

Disqualification.

Art. 58.—If a competitor is disqualified in a race he loses all right to a prize.

Official and Public Reprimand.

Art. 59.—A public and official reprimand is pronounced by the stewards or by the Sporting Committee of the Automobile Club of France, and involves the insertion in a public journal of an official notification by the Sporting Committee.

Fines.

Art. 60.—The moneys received in fines shall be paid into the funds of the Sporting Committee, to be distributed or devoted to sporting competitions.

II.—Regulations as to Records.

Timekeepers.

Article 1.—The Sporting Committee shall appoint the official timekeepers and shall prepare a list of them every year.

Art 2.—Timekeepers to be eligible for appointment must (1) possess a reliable chronometer stop-watch, certified as firstclass by the Observatories of Besançon, Geneva, or Kew; (2) furnish the name of the maker of their chronometer stop-watch.

Art. 3.—The Sporting Committee may, when it sees fit, require the timekeepers to renew the certificates as to their chronometers being first-class. Certificates must be renewed every three years.

Art. 4.—The appointment of timekeepers is revocable at any time. Before appointment they must: (1) Submit to an examination permitting of the chronometrical test, (a) of 10 tests of from 500 metres (500 yards) and under; (b) of 10 tests of from 500 metres (yards) to 2,000 metres (yards); (c) of two tests of 20 kiloms. (15 miles) at least, or a test of 50 kiloms. (38 miles), the stopwatch showing the time of each lap and the time of the total distance.

In the above tests the candidate for appointment as timekeeper shall write down on the forms, of which a model is deposited at the offices of the Automobile Club of France, the times recorded by him. At the same time a certified official timekeeper shall make similar entries, but independently of the candidate. The candidate shall remit these forms, duly filled up, to the certified timekeeper in a sealed envelope. At the end of the tests the certified official timekeeper shall forward these forms to the Sporting Committee of the Automobile Club of France, together with the results of his own checking, certifying that the examination has been properly conducted, and that there has been no collusion, comparison, or correction of results.

Art. 5.—The Sporting Committee shall decide on the appointment after examining and comparing the written results. A candidate who has been rejected may re-enter for election after a month.

Art. 6.—Timekeepers must sign forms recording the times taken by them. Any timekeeper signing a record not made by himself will be ipso facto disqualified. He will also be disqualified by the simple decision of the Sporting Committee that his records have not been confirmed.

Art. 7.—The Sporting Committee takes cognizance of records on the track and road records. Each of these two categories comprises records of distance and time, as well as the records for both categories defined by Article 9 of the Racing Regulations.

Art. 8.—The distances officially recognized for record racing are: On the track, 500 metres (500 yards); from 1 to 100 kiloms., per kilom. (1,094 yards); and for distances beyond 100 kiloms., per 50 kiloms. On the road: 500 metres; from 1 to 10 kiloms., by kiloms.; from 10 to 50 kiloms., by 10 kiloms.; from 100 kiloms., by 100 kiloms. The official distances in English miles: Distances of miles, 50 miles and 100 miles will be recognized.

Art. 9.—All races for records must be made from standstill, and vehicles must be started only with their own power.

Art. 10.—Races for records of 500 metres (541 yards) and of from 1 to 10 kiloms., inclusive, may be made by flying start.

Art. II.—The time records of the Automobile Club of France are records by the hour without limit

Art. 12.—The time records from town to town are also by the hour without limit (homologous).

Art. 13.—No record will be recognized as official unless it has been established over distances rigorously tested, and unless the time has been checked by several official timekeepers recognized by the Automobile Club of France.

Track Records.

Art. 14.—Starts for track records shall take place from a tape.

Art 15.—Attacks on the record shall be timed according to the laps round the track and by the hour up to 100 kiloms., by kilom., and by hour up to 200 kiloms., and by the 5 kiloms., and by the hour from 200 kiloms. upwards.

Whilst timing records timekeepers are expressly advised to take the times of the English distances at the half mile, mile, and all the military distances, especially the 10, 20, 30, 40, 50, and 100 miles, and above the last-named distance by the 100 miles.

Road Records.

Art. 16.—Road records straight ahead are recognized from 1 to 50 kiloms., above that distance they are taken by the 50 kiloms. Road records permit of embracing the outward and return journey for all distances.

Timekeepers' Fees.

Art 17.—Timekeepers are forbidden to accept any remuneration over and above the tariff fixed below, viz.: For a day, or part of a day, occupied in racing or in getting to and from a race, 30 francs (25s.)

Art. 18.—The traveling expenses of timekeepers are arranged by mutual consent.

Art. 19.—Every timekeeper must, at his own expense and on his own responsibility, procure such assistance as he may require in working out his calculations, or for any other outside act or operation required, not strictly coming within the province of a timekeeper.

Art. 20.—Timekeepers may be temporarily suspended or have their appointment revoked for any act affecting their private or professional honor. Provided, that this step cannot be taken unless by order of the Sporting Committee, after the accused timekeeper has been heard.

Art 21.—No timekeeper shall be required to act as such for more than six hours at a stretch.

Operation of the Regulations.

Special Article.—The present regulations shall come into force and be binding on all promoters of Automobile races beginning January 1, 1900.

APPENDIX

TYPES OF MODERN VEHICLES

ROAD WAGON

Total weight of this vehicle is 1,000 pounds, the batteries of which weigh 450 pounds. The horse-power of the motor is nominally 2½ horse-power, but is capable of being worked to 5 horse-power for a short period of time.

Its maximum speed is twelve miles per hour; its mileage capacity is twenty-five miles on one charge of the batteries on a hard, level road; and it has successfully mounted 14-per cent grades, carrying two people.

The number of battery cells in the vehicle is thirty-six. Average current consumption at maximum speed is 14 amperes. The batteries have a capacity that will maintain this discharge for two and one-half consecutive hours.

There is but one motor, which drives the rear axle and this directly through a differential gear. It has three gradations of speed, viz., three, six and twelve miles per hour. The vehicle is equipped with ball bearings throughout, and runs on four wheels of an equal diameter of thirty-six inches. The length of the box over all is sixty-two inches, with a width of twenty-two and a half inches and a seat width of thirty-six inches. It is built with or without a top.

STANHOPE

Total weight of this vehicle is 2,200 pounds, the batteries of which weigh 900 pounds. The horse-power of the motors is nominally 5 horse-power, but is capable of being worked up to 10 horse-power for a short period of time.

Its maximum speed is twelve miles per hour, and its mileage capacity is thirty-five miles on one charge of the batteries on a hard, level road. It has mounted successfully 12 and 14 per cent grades carrying two people.

The number of battery cells in this vehicle is forty. The average current consumption on a hard, level road at maximum speed is twenty-two amperes and the batteries have a capacity that will maintain this discharge for three consecutive hours.

There are two motors, which drive the rear wheels independently of one another, and the vehicle has four gradations of speed, viz., five, seven and a half, ten and twelve miles per hour. It is equipped with ball-bearing axles throughout, runs on wheels of forty inches diameter rear, thirty-four inches diameter front, and has deep, full-width carriage seat

VICTORIA

Total weight of this vehicle is 2,400 pounds, batteries of which weigh 900 pounds. The horse-power of the motors is identical with those of the Stanhope.

Its maximum speed is twelve miles per hour and its mileage capacity, carrying four people, is about thirty miles on one charge of the batteries. The number of battery cells is the same as in the Stanhope and average current consumption is twenty-six amperes on a hard, level road at maximum speed.

The motors are connected with the rear wheels the same as in the Stanhope and it has the same gradations of speed. Also the same size of wheels.

DELIVERY WAGON

Total weight of this vehicle is 3,000 pounds, the batteries of which weigh 1,250 pounds. The horse-power of the motors is nominally 7 horse-power, but is capable of being worked up as high as 14 horse-power for a short period of time.

Its maximum speed is ten miles per hour and it has a mileage capacity of thirty miles on one charge of the batteries on a hard, level road.

This vehicle, in the service of the company for which it was built, has been in nearly all the principal cities of the United States and successfully mounted all the grades in those cities, ranging from 2 to 14 per cent.

The actual test of this vehicle when operated by the express company gave a mileage of forty-five miles on one charge of the batteries; and after being in use eight months, another test made by this company gave a mileage capacity of thirty-eight miles on one charge of the batteries. The average current consumption of the vehicle with two men on it on a hard, level road is thirty-three amperes and with 1,200 pounds additional load, about ten amperes more. The batteries will maintain a discharge of forty amperes for three consecutive hours.

There are two motors, one attached independently to either rear wheel. It has three gradations of speed, two and one-half, five and ten miles per hour. It is equipped with ball bearings throughout, and runs on wheels of thirty-eight inches diameter rear and thirty-two inches diameter front. It has a four-foot eight-inch tread and the box is nine feet two inches over all in length.

ELECTRIC TRUCK

This is one of the most successful electric trucks that has ever been constructed and put on the market. The total weight of this vehicle is 3,800 pounds, the batteries of which weigh 1,680 pounds.

The horse-power of the motors is nominally 9 horse-power, but they are capable of being worked to 18 horse-power for a short period of time.

Its maximum speed is eight miles per hour, and its mileage capacity without load is thirty miles on one charge of the batteries, on a hard, level road. The number of battery cells in the vehicle is forty, and its average current consumption at third speed when empty with two men on the box is thirty-seven amperes. With a load of 2,500 pounds additional, its current consumption is sixty amperes on a hard, level road. The batteries have a capacity that will maintain this discharge for three hours consecutively or twenty-five miles with load. There are two

124

motors, one connected independently to either rear wheel, and it has three gradations of speed, viz., two, four and eight miles per hour.

This is also equipped with ball bearings throughout and runs on wheels of thirty-eight inches diameter rear and thirty-two inches diameter front. The length of the body over all is eleven feet four inches, and the width between side bars is three feet ten inches.

CHARGING OUTFITS

While charging outfits are subject to modification in their arrangement and in the makes of instruments used, Plate VIII shows this so completely in principle that the modus operandi shotdd always be the same in all.

The main line wires tap on to the fuse block (marked fuse), the purpose of which is to protect the wagon from any damage should too great an amount of current be absorbed by it, as well as to protect the machinery and accessories that supply the current From this fuse block the wires go to a switch which is used for breaking the circuit open that supplies the current.

On the right-hand side is an ampere meter which is wired into one side of the main wires that lead to the charging plug and which registers the amount of current flowing; the other main wire from the switch, passing through the rheostat (which has several points of contact), is used to regulate the amount of current flowing, so that the vehicle may be charged in a short time, over night or in any intermediate time that it is desired to charge it, the amount of current desired being wholly controlled by the rheostat.

On the left-hand side is a volt meter tapped on to the main lines, which will indicate the voltage of the main circuit when the wagon is charging and, by opening the main switch when the wagon is supposed to be charged, the voltage of the batteries can be read from the same instrument.

Another pair of wires, usually twisted together in the shape of a cord, extends from the switch board to the charging plug, which has, as shown, a receptacle provided on the under side of the vehicle for its insertion.

When it is desired to charge a wagon, it should be first noted that the rheostat is at the side which will allow the least amount of current to flow. The main switch should then be opened, the controller of the vehicle adjusted for charging and the charging plug inserted in the receptacle provided for it on the vehicle; after which the switch that closes the circuit on the switch board should be thrown in and the amount of current it is desired to use in the vehicle regulated by the rheostat.

ELECTRIC MOTORS

Plate IX shows one form of electric motor designed for carriage use, completely assembled except the cover that goes over the brushes and commutator. The object of the photograph is to illustrate the method of applying a band brake.

126

A pulley is placed on the end of the motor shaft, around which a steel strap is placed, and the lever top is used to actuate this brake by being connected with some form of lever at the driver's seat. On the other end of this shaft the pinion is placed which drives the main gears of the wagon; therefore, by this method, a very forcible acting brake is obtained with the minimum amount of exertion on the part of the operator.

Plate X shows this same form of motor disassembled. The armature, or revolving part of the motor, i. e., the part that actually sets the wheels in motion, is resting on top of the field frame, showing fully the way the wires look after being placed in the armature and connected up to the commutator at one end. This also shows the location of the ball bearings of the motor in the journal and indicates very clearly the general constructional features of a motor and where to proceed to disassemble it

Plate II represents a sectional view of this same motor, showing the location of the band brake, the gear wheel, the coils of wire in the armature where they are connected with the commutator, and is especially drawn to show the location of the ball bearings and their method of adjustment.

It will be noticed that on the pinion end, the ball bearings are mounted directly on the shaft and are adjusted by two nuts threaded upon it, which are accessible from the outside. On the commutator end —the end where the band brake is—a different arrangement exists. The heat of an electric motor is often considerable, and as the contraction and expansion of the shaft will not permit a regular form of ball bearing application this contraction and expansion have to be allowed for, so that they may not interfere with the adjustment.

On this end the ball bearings are mounted on the steel sleeve through which the shaft runs and on this sleeve the adjusting nuts are placed on the outside. In this way, adjustment of the ball bearings being once made, it is practically permanent and no contraction or expansion of the shaft will interfere with it.

Plates XII and XIX show cross section drawing and photograph reproduction of a differential gear. In Plate XIX we see two beveled gears carrying between them two pinions, on the face of which is mounted the gear proper which drives the vehicle.

In Plate XIX this is represented as one beveled gear, being fastened on a shaft which extends into another shaft or sleeve on which the other gear is mounted.

By this means when a vehicle is turning around and one wheel is running faster than the other it automatically adjusts itself to any difference in speed by means of the small pinion between the beveled gears, as shown in Plate XIX. This is only one form of differential gear, but it has in it the principle of any differential gear that may be used, and which it is absolutely necessary to use with any vehicle having but one motor to drive two wheels. Otherwise it would be practically impossible to turn the vehicle around.

Plate XIII shows drawing of one rear wheel and one motor, on a vehicle having two motors to propel it, arranged so that one motor drives either wheel of the vehicle independently of the other motor.

This also shows a steel hub with ball bearings mounted in it on the axle, which is stationary, and on this hub a large gear wheel is mounted which is operated by the small pinion on the end of the motor shaft. It also shows the position that the band brake occupies in a vehicle of this kind; as there, are of course, two brakes, one on each motor.

The advantages of the double motor equipment are many, especially for heavy work. When motors are properly designed, if an accident should occur to either motor, the other motor ought to be able to run the vehicle.

This drawing also illustrates the compression cup on the forward side of the motor bearings by which the ball bearings of the motor are lubricated.

To remove the wheel of the vehicle from the axle, nothing more is required than for removing an ordinary wheel. There is simply a dust cap to take off, which screws into the hub on the front side; then the nut from the end axle; after which the wheel can be removed without disturbing any adjustments whatever.

The object of using a steel hub is in order that the gear wheel may be mounted on the butt of the hub in a very secure way and also to make it an easy matter to slip on a new gear wheel in case of accident or necessity from long wear.

Plate XIV shows controller with the handle removed in such a position that the three different contacts made in its rotation for three different speeds are easily seen; also the eight receptacles on the upper side which receive the wires from the batteries are indicated by little round rings as shown in Plates XV, XVI, XVII and XVIII.

Plate XVIII differs from the other diagrams inasmuch as in lieu of putting the controller on third speed to charge, the controller handle, when brought back to neutral line for stopping the vehicle, is placed in position to charge.

The difference in the diagrams is easily traced out by referring to point "A" and "B" in Diagram XVIII and comparing them with the other diagrams.

PLATES

PLATE I
DR. CHURCH'S STEAM ROAD VEHICLE, 1834

PLATE II

OLD ENGLISH STEAM MOTOR AND COACH, ABOUT 1815

PLATE III
SQUIRE STEAM CARRIAGE, 1833

PLATE IV
ELECTRIC STANHOPE

PLATE V
ELECTRIC ROAD WAGON

PLATE VI
ELECTRIC VICTORIA

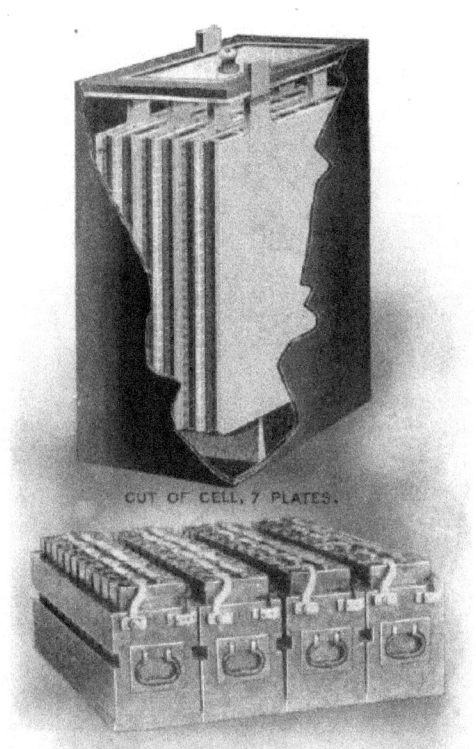

CUT OF CELL, 7 PLATES.

40 BATTERIES IN TRAYS OF 10 EACH.

PLATE VII
BATTERY CELL. 40 BATTERIES IN TRAYS.

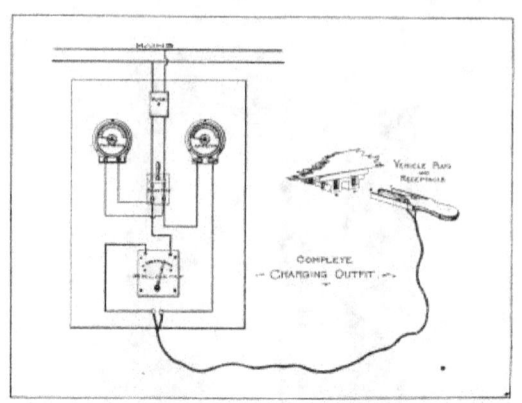

PLATE VIII
BATTERY CELL. 40 BATTERIES IN TRAYS.

PLATE IX
ELECTRIC MOTOR, ASSEMBLED

136

PLATE X
ELECTRIC MOTOR DISASSEMBLED

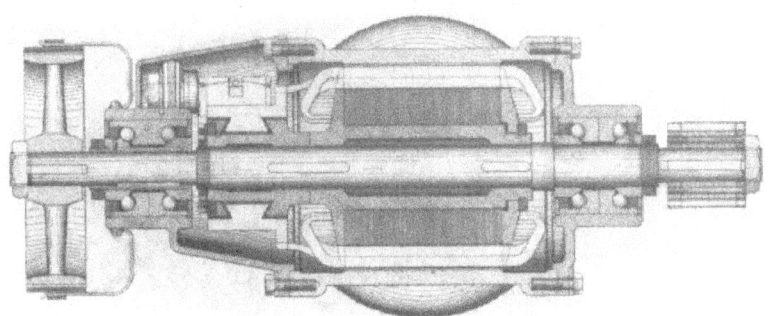

PLATE XI
ELECTRIC MOTOR, CONSTRUCTIONAL VIEW

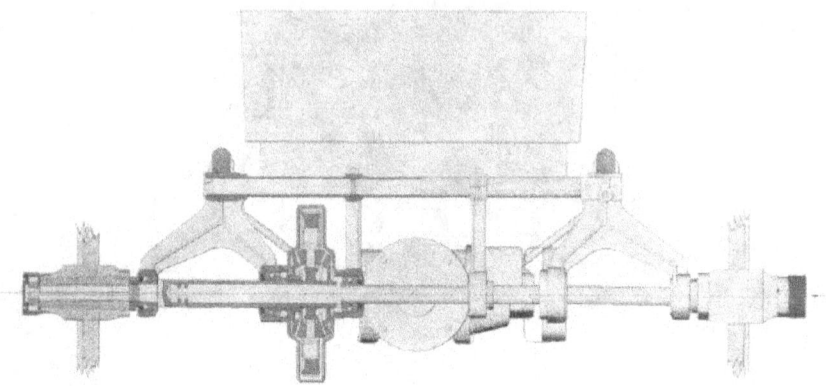

PLATE XII

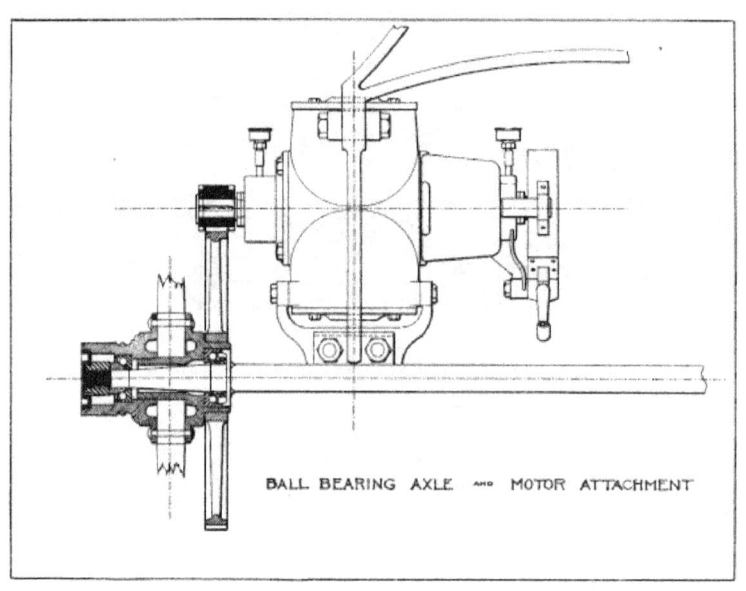

BALL BEARING AXLE ᴀɴᴅ MOTOR ATTACHMENT

PLATE XIII

138

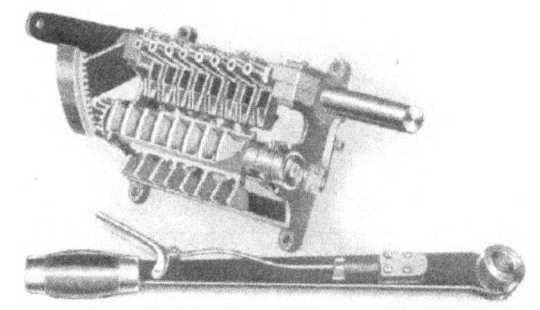

PLATE XIV

CONTROLLER SHOWING CONTACTS

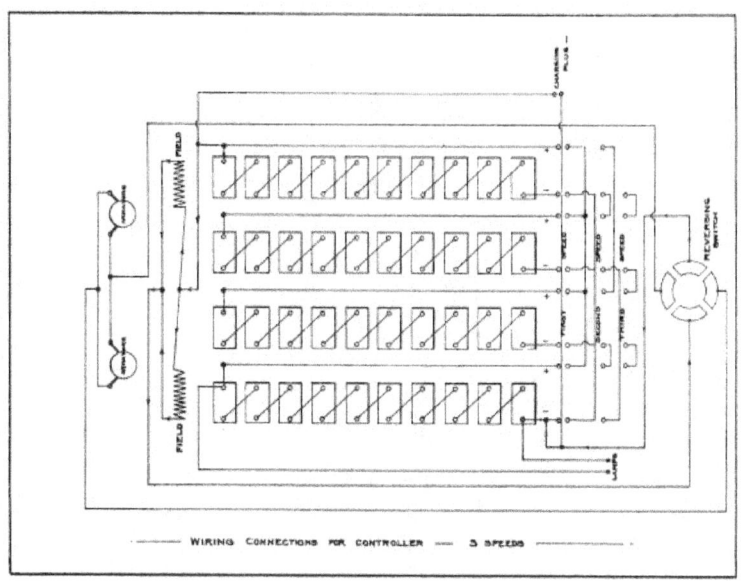

PLATE XV

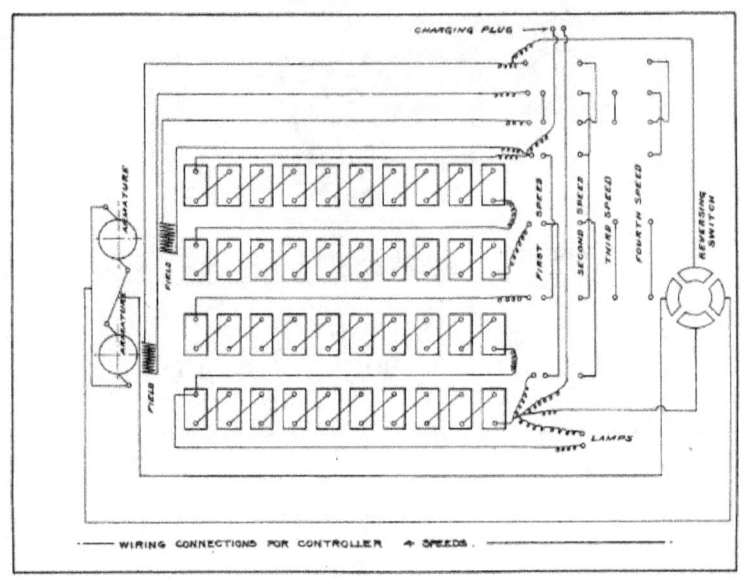

PLATE XVI

140

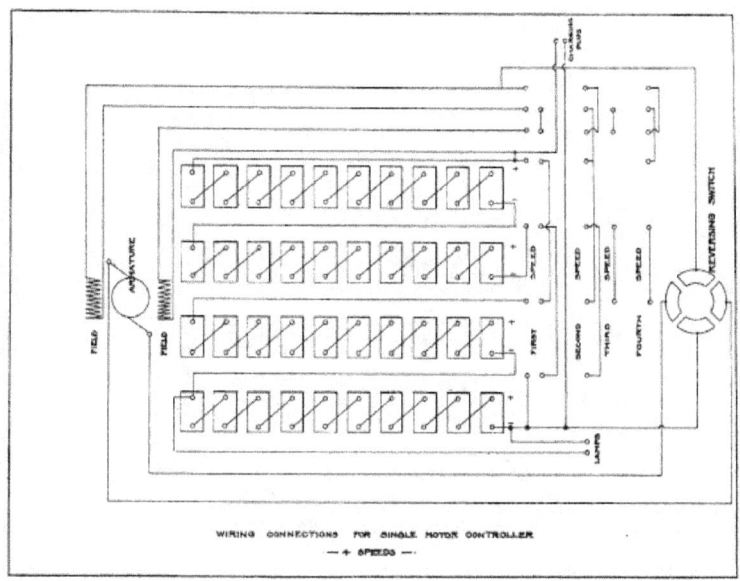

WIRING CONNECTIONS FOR SINGLE MOTOR CONTROLLER
— 4 SPEEDS —

PLATE XVII

141

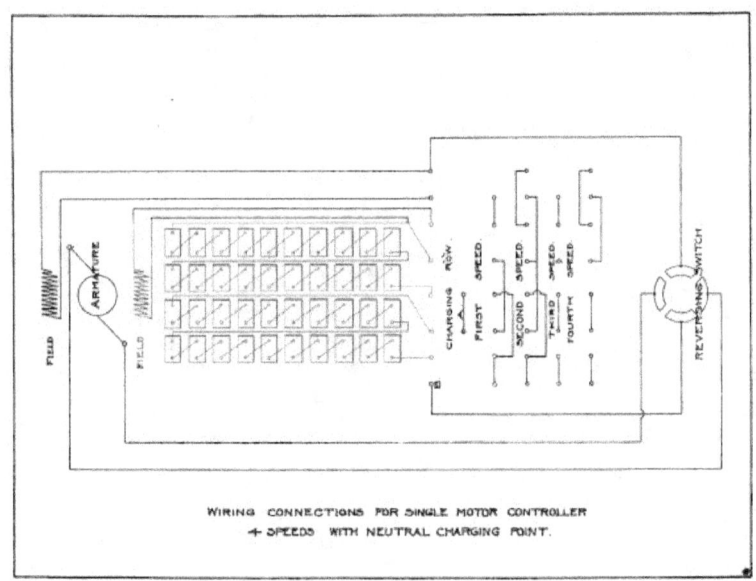

WIRING CONNECTIONS FOR SINGLE MOTOR CONTROLLER
4 SPEEDS WITH NEUTRAL CHARGING POINT.

PLATE XVIII

PLATE XIX

142

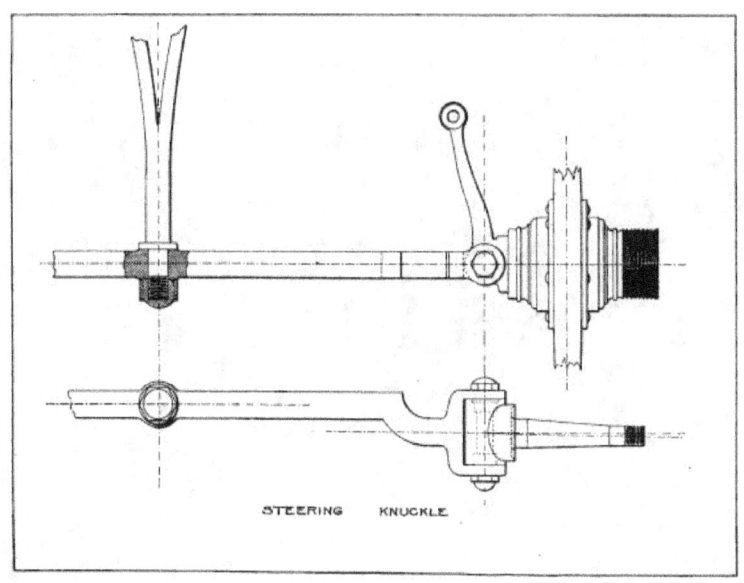

STEERING KNUCKLE

PLATE XX

143

PLATE XXI
ELECTRIC DELIVERY WAGON

PLATE XXII
ELECTRIC TRUCK

www.ingramcontent.com/pod-product-compliance
Lightning Source LLC
Chambersburg PA
CBHW070251190526
45169CB00001B/363